William Marcet

On Chronic Alcoholic Intoxication

Alcoholic Stimulants in Connection with the Nervous System

William Marcet

On Chronic Alcoholic Intoxication
Alcoholic Stimulants in Connection with the Nervous System

ISBN/EAN: 9783744670319

Printed in Europe, USA, Canada, Australia, Japan

Cover: Foto ©berggeist007 / pixelio.de

More available books at **www.hansebooks.com**

ON

CHRONIC ALCOHOLIC

INTOXICATION,

OR

ALCOHOLIC STIMULANTS IN CONNEXION
WITH THE NERVOUS SYSTEM;

WITH A

SYNOPTICAL TABLE OF CASES.

BY

W. MARCET, M.D., F.R.S.,

FELLOW OF THE ROYAL COLLEGE OF PHYSICIANS; ASSISTANT-PHYSICIAN TO THE
WESTMINSTER HOSPITAL, ETC., ETC.

LONDON:
JOHN CHURCHILL, NEW BURLINGTON STREET.
MDCCCLX.

J. E. ADLARD PRINTER, BARTHOLOMEW CLOSE, E.C.

INTRODUCTION.

THE therapeutical properties of zinc having become during the last few years a subject of much medical interest, I was induced to commence in 1855 a series of comparative inquiries respecting the action of oxide of zinc in epilepsy, chorea, mild hysteria, paralysis and lead palsy, cases of exhaustion from excessive mental exertions, and in another peculiar chronic disorder of the nervous system which is the subject of the present treatise. It soon became obvious that in certain of the above diseases the treatment adopted effected a complete recovery, or was attended by a very marked improvement, that in others it was followed by no material benefit, and even in a few instances of hysteria was open to objection. I observed that this substance was most remarkably beneficial when given to patients suffering from a chronic disorder of the nervous system, characterised by sleeplessness, giddiness, headache, flying specks passing before the eyes (muscæ volitantes), noises in the ears (tinnitus aurium), hallucinations, trembling, and want of co-ordination of the voluntary motions; the disease being frequently accompanied by a morbid condition of the organs of digestion. It then appeared to me probable, and I soon afterwards discovered, that these symptoms were owing to one and the same cause, viz., *the excessive use of alcoholic stimulants.*

I now commenced a series of investigations as to the property of oxide of zinc in controlling and curing the disorder in question, taking notes of the symptoms and other particulars of the cases of this affection admitted under my care at the Westminster Hospital. In January, of the present year, I communicated to the London Western Medical Society a paper, showing that oxide of zinc was exceedingly efficacious in the treatment of chronic alcoholic intoxication; and abstracts of this paper were published shortly afterwards in the 'Medical Times and Gazette,'[1] and in the 'Lancet.'[2]

Subsequent observations having fully confirmed the correctness of the above-mentioned results, I have thought it might be useful to publish in its present form an account of the chronic functional disturbance of the nervous system brought on by the abuse of spirituous stimulants, and of the treatment adopted in these cases.

I was engaged in the preparation of this treatise for the press, when the valuable publication on chronic alcoholism, by Magnus Huss of Stockholm, came under my notice. In this work the author has fully and accurately described the disease now under consideration. It may be observed, however, that his attention has been more particularly directed to the symptoms and pathology of the disorder, while in the present volume comparatively great stress has been laid on the predisposing and immediate causes of the illness, and especially

[1] 12th February, 1859.
[2] 2d April, 1859.

on its mode of treatment. In order to be assured of the correctness of my statements, I have confined myself almost exclusively to the results of my own experience, avoiding as much as possible filling up omissions by borrowing from the writings of others. The following, therefore, is a mere sketch of a vast and important subject; but, however incomplete it may be, it will, I trust, be found not altogether devoid of interest and usefulness.

The reader will not meet in the present work with any account of the physiological properties of alcohol, and is referred for information on this subject to Dr. Carpenter's treatise on the use and abuse of alcoholic liquors.[1]

[1] See also a paper communicated by the author in 1859 to the British Association for the Advancement of Science, on 'An Experimental Inquiry into the action of Alcohol on the Nervous System.'

36, CHAPEL STREET, BELGRAVE SQUARE.

CONTENTS.

	PAGE
Definition of chronic alcholism	2
Difference between chronic alcoholism and delirium tremens	3
Division of the subject	4

SYMPTOMS OF CHRONIC ALCOHOLISM . . 5
    Inability to sleep 9
    Giddiness, headache, hallucinations . . 11
    Weakness 14
    Difficulty of breathing . . . 15

CAUSES PREDISPOSING TO CHRONIC ALCOHOLISM . 17
    Influence of quality of spirituous stimulants taken 18
    ,, quantity ditto ditto . 22
    ,, period during which the habit of excessive drinking has been indulged . 26
    ,, age 27
    ,, sex 30
    ,, temperament . . . 31
    ,, habits; smoking tobacco . . 32

	PAGE
Influence of profession	38
„ circumstances connected with the food taken	42

IMMEDIATE CAUSES OF AN ATTACK OF CHRONIC ALCOHOLIC INTOXICATION . . 46

An exceptional excess	47
Coexisting disease	48
Cerebral concussion	55

AFFECTIONS OF THE NERVOUS SYSTEM RESEMBLING CHRONIC ALCOHOLIC INTOXICATION . 57

| Nervous disorder owing to mental exhaustion | ib. |
| „ „ violent emotions | 61 |

PROGNOSIS 62

TREATMENT OF CHRONIC ALCOHOLISM . . 64

Inveterate nature of the habit of drinking	65
Means of checking the habit of drinking	66
Importance for the physician of obtaining the full confidence of his patient	ib.
The cause of intemperance to be removed	67
Patients not inveterate drinkers	69
Diet to be prescribed	71
Substitutes for alcoholic stimulants	73
Treatment adopted by Magnus Huss in cases of chronic alcoholism	74
Principle of the treatment adopted by the author	77

CONTENTS. xi

	PAGE
PHYSIOLOGICAL PROPERTIES OF OXIDE OF ZINC	78
Results of Dr. Herpin's investigations	79
„ the author's ditto	81
Nausea and sickness produced by oxide of zinc	82
Oxide of zinc objectionable in certain cases	84
Sleep, an effect produced by oxide of zinc	85
Oxide of zinc, not a slow poison	87
THERAPEUTICAL PROPERTIES OF OXIDE OF ZINC	91
Dr. Golding Bird's opinion on the action of zinc	92
Case reported of a dog treated with oxide of zinc for a convulsive disorder	ib.
Action of oxide of zinc in various disorders of the nervous system	94
Effect of oxide of zinc in a case of gastralgia	100
CHRONIC ALCOHOLISM TREATED WITH OXIDE OF ZINC	102
Circumstances liable to modify the action of oxide of zinc in cases of chronic alcoholism	ib.
Usual effects of oxide of zinc in simple cases of chronic alcoholism	104
Classification of cases selected in order to illustrate the therapeutical action of oxide of zinc in cases of chronic alcoholism	105

CLASS I.—DIVISION I.

Cases of chronic alcoholism not complicated by other diseases, and occurring during the period of intemperance 106

CLASS I.—DIVISION II.

Cases of chronic alcoholism not complicated by other diseases, and occurring in patients having resumed habits of sobriety 120

CLASS II.—DIVISION I.

Cases of chronic alcoholism, accompanied by other diseases, and occurring during the period of excessive indulgence 129

CLASS II.—DIVISION II.

Cases of chronic alcoholism complicated by other diseases, and occurring after the habit of excessive drinking had been given up . . . 136

OBSERVATIONS ON SYNOPTICAL TABLE . . 140
SYNOPTICAL TABLE 148

APPENDIX 171

ALCOHOLIC STIMULANTS

IN CONNEXION WITH

THE NERVOUS SYSTEM.

The injurious effects upon health, arising from the abuse of wine, beer, or spirits, and generally from all alcoholic stimulants, have been usually considered under the two following heads: *First*, as the immediate consequence of an excessive indulgence on a particular occasion; and *secondly*, as the result of long-continued intemperance. It is not my intention to dwell upon the immediate action of alcohol upon the human body, producing drunkenness; this subject having been admirably treated by Drs. Trotter,[1] Macnish,[2] Roesch,[3] and more

---

[1] 'An Essay, Medical, Philosophical, and Chemical, on Drunkenness,' 1804.
[2] 'The Anatomy of Drunkenness,' 1832.
[3] 'De l'abus des boissons Spiritueuses,' 1839.

recently by Dr. Carpenter in his valuable Prize Essay 'On the Use and Abuse of Alcoholic Liquors in Health and Disease.' I purpose to direct the reader's attention to a most distressing form of disease affecting the nervous system, which those unfortunate persons who persist in drinking to excess seldom escape, and which constitutes a state of prolonged or chronic poisoning. This affection is known by the name of *chronic alcoholic intoxication*, or *chronic alcoholism*, and is thus defined by Magnus Huss: "The name *chronic alcoholism* applies to the collective symptoms of a disordered condition of the mental, motor, and sensory functions of the nervous system, these symptoms assuming a chronic form, and without their being immediately connected with any of those (organic) modifications of the central or peripheric portions of the nervous system which may be detected during life, or discovered after death by ocular inspection; such symptoms, moreover, affecting individuals who have persisted for a considerable length of time in the abuse of alcoholic liquors."

The habit of indulging too freely in spirituous

beverages, even without their producing intoxication, is often attended at first with no apparent evil result, and there is little or no warning given of the injury done to the constitution; but sooner or later, the injurious effects of excessive and frequent alcoholic libations will become obvious, occurring sometimes under the form of delirium tremens, although much more frequently assuming the condition of the chronic disease which will be subsequently described.

The difference between *delirium tremens* and *chronic alcoholism* is distinctly marked: the former, consisting of an acute and violent disturbance of the functions of the nervous system, lasting generally, according to Macnish, from four to ten days; whilst the latter, although resembling the other in a mitigated condition, assumes the form of a protracted illness, with none of the paroxysms of violent delirium so peculiar to delirium tremens. Chronic alcoholic intoxication, again, is a state of long and uninterrupted suffering, allowing the patient no rest day or night. He may have given up the habit of drinking before the outbreak of the disease; or if attacked during a period of excessive indulgence, he may endeavour by a great effort to shake off his old and pernicious habit, but, even if successful, he will not unfrequently be disappointed in the hope of regaining his lost health. Week after week, month after month, year after year, he patiently waits for the termination of his sufferings, and endeavours to apply the remains of his strength to such occupations as will tear him away for the time from his miseries, although usually he finds himself so weak that he is denied even this poor source of relief.

My attention having been directed for several years to an inquiry into the symptoms and treatment of the disorder in question, I have decided upon communicating, in this little volume, the results of those inquiries to the Medical profession and the Public.

I shall begin with a description of the *symptoms* of chronic alcoholic intoxication. The causes, predisposing the individual to suffer from the disease, will next be duly investigated; to be followed by an inquiry into the immediate causes of the attack, and a short account of certain diseases resembling chronic alcoholism, although not depending on excesses in spirituous drinks. I shall afterwards insist more particularly on the treatment of the disorder, and show how effectually it may be placed under medical control; concluding by a faithful report of the cases which have fallen under my care; the notes of which have been taken during the patient's visits, and consequently not depending for accuracy upon mere recollection.[1]

SYMPTOMS OF CHRONIC ALCOHOLISM.

Dr. Carpenter, in his Prize Essay (p. 30), alludes to the symptoms of chronic alcoholism in the following words: " It is important to remark that a slighter form of this disorder (delirium tremens), marked by tremors of the hands and feet, deficiency of nervous power, and occasional illusions, will sometimes occur as a consequence of habitual tippling, even without

---

[1] A synoptical table is appended to this treatise, giving an account of forty-eight cases of chronic alcoholism treated by the author at the Westminster Hospital.

intoxication having been once produced. And a still slighter manifestation of the want of control over the muscular apparatus, the trembling of the hands in the execution of a voluntary movement, is familiar to every one as extremely frequent among the habitually intemperate." And Dr. Carpenter further observes (p. 46), " That the effects of drunkenness are highly inimical to a permanent healthy state of the brain, is often proved at a great distance of time from the course of intemperance, and long after the adoption of regular habits."

The remote effects of alcoholic poisoning have been also very ably considered by a French author—Roesch; but he includes *chronic alcoholism* among a number of other diseases resulting indirectly from long-continued intemperance, and thus omits a classification which is of the highest importance respecting the treatment of the disorder in question.

The symptoms of the disease depend on a functional disturbance of the properties of the nervous system, which may last for weeks, months, or years, even after the habit of excessive drinking has been given up.

On first applying to his medical adviser, the patient will probably not state the cause of his illness, and thus seriously mislead the physician in his estimation of the nature of the complaint. If we try to account for this difficulty of establishing the cause of the disease in cases of chronic alcoholism, it will be found that in some instances the patient is ashamed of his intemperance, and will not confess it. In others, he considers that the nature of his occupation is such as to require an excessive amount of drink; he is seldom or never drunk, in his opinion he takes no more than is absolutely required, and he is not aware of his suffering from alcoholic stimulants. Some will positively disbelieve that their illness can be owing to the abuse of alcoholic liquors, as they have been under a pledge to drink very little or none at all for some time previously; but it will be noticed, in the course of the examination, that before taking the pledge these individuals were thorough drunkards, and had been obliged to give up drinking on account of their health. Finally, in those instances where the mind has been affected through frequent fits of drunkenness and repeated attacks of delirium tremens, the patient may have great aversion from giving plain answers to the questions of the physician, and thus lead him to understand that he never indulged too freely in spirituous drinks.

There is something peculiar in the look and gait of individuals in the habit of drinking to excess, or even of habitual tipplers, which will greatly assist in discovering the nature of the complaint, even before addressing the patient. His peculiar complexion, often sharp features, or, if he be fat, the injected cheeks and nose, and their violet appearance, the trembling of the limbs, often of the whole body, or a want of steadiness and co-ordination in the movements, not very unlike incipient chorea—all these are so many symptoms that the medical practitioner will not fail to observe. On conversing with such patients their intellect will not often be found blunted, and the account they give of their sufferings is perhaps remarkably clear. It may be, however, that the patient has fallen into a state of melancholia, and fancies his doctor is attempting to do him some injury, in which case the sufferer will endeavour to turn aside the conversation, and adroitly avoid the subject. I have observed a well-marked instance of this kind in a boy of eighteen, whose case I had great difficulty at first in making out; but a fortnight afterwards, his health having much improved, he gave me a clear account of his illness. Another patient, although he had given up the habit of drinking to excess, taking no more on an average than two pints of beer daily for the last six years, yet assured me that he was occasionally unconscious of what he was doing, and that his friends sometimes thought he was insane.

Want of Sleep.—One of the prominent symptoms of chronic alcoholism is want of sleep and great restlessness at night; the sufferer perpetually keeps turning over in bed, and, as soon as he shuts his eyes, extraordinary visions, mostly of a painful kind, appear before him. For example, a patient told me he frequently saw a funeral passing as he was endeavouring to compose himself. Sleep also, when obtained, is disturbed by frightful dreams, which appear often to indicate a considerable degree of mental excitement; the patient frequently dreaming that he has been at his work all night; he awakes in the morning exhausted, and almost incapable of any exertion. The ill effect produced on sleep by excessive drinking is well illustrated in the following case of A. T——, aged 25 (Case 45). He is in the habit of taking daily one pint of beer and three or four glasses of gin. Once a week—on Saturday, he increases his allowance to seven or eight pints of beer and five or six glasses of gin; he cannot sleep on the nights of Saturday, Sunday, Monday, and Tuesday, but on Wednesday night he sleeps better, and pretty well on Friday night. He is also troubled with hallucinations and the other usual symptoms of chronic alcoholism, being invariably worse after his Saturday's excesses.

Trembling.—In the day-time the patient is seized with trembling, especially when in the sitting posture; some hardly tremble at all when walking. The trembling may be very slight indeed, or confined to a particular part of the body, being frequently visible in the tongue, or it may only occur at intervals; some merely tremble in the morning on getting up, and many affected by chronic alcoholism experience more or less difficulty in dressing themselves from want of sufficient control over their movements. As already mentioned, there is often much awkwardness in the performance of voluntary motion, even when the body is not subject to a conspicuous trembling; thus, in the act of drinking, a liquid will be spilt from the cup; or a light will be put out instead of being snuffed. It is remarkable how long this condition may last, and how rapidly it disappears under an appropriate treatment.

Giddiness and Headache.—The patient also generally complains of great giddiness, more especially when suddenly raising his head from the recumbent posture : headache and a ringing noise in the ears (*tinnitus aurium*) are frequent, although not constant, symptoms.

Hallucinations.—The occurrence of hallucinations, so characteristic of delirium tremens, is very often noticed in chronic alcoholism, although usually to a much more subdued extent. They mostly affect the organs of sight and of hearing. For instance, one of my patients, when walking in the street, had seen ropes dangling about his head; to another, objects appeared as if they were double; some perceived occasionally insects creeping about: the various visions often disappearing as soon as the attention was directed to them. These factitious perceptions of the sight appear sometimes so real that the individual moves aside to avoid an imaginary object standing in his way. A cabman (Case 47) I was treating for chronic alcoholism told me he frequently pulled up his horse suddenly, or drove to one side of the street, lest he should run over some obstacle he distinctly saw in front of his horse, and which he afterwards found not to exist in reality. In his case objects appeared to be multiplied to as many as ten times their real number, so that if a lamp-post, a man, or a cart, happened to be near him, he perceived ten lamp-posts, or ten men, or as many carts. He could not possibly make out which object was really to be avoided, and was obliged to give up driving on account of the risk of an accident. In most cases the patient is occasionally, or perhaps constantly, troubled with shadows or a black mist or flying specks (*muscæ volitantes*) passing rapidly before his eyes, and causing a dimness of sight, especially when he is looking attentively at something; in the act of reading, for example, the book is suddenly darkened, and a state of almost complete blindness ensues, lasting a few minutes. I have met with one instance where the patient perceived spots of all kinds of colour.[1] During the long and sleepless nights, aberrations of the sight frequently happen. The wife of a patient I was treating for chronic alcoholism told me her husband often fancied, whilst lying awake, that he saw rats and cats, and various other descriptions of animals, on the bedclothes; he used to doze at intervals, and in the morning could not remember anything of the nightly visions. The aberrations of the sense of hearing are not so frequent, but I have met

[1] Magnus Huss has observed cases of chronic alcoholism where objects appeared peculiarly coloured. He reports having met with two instances of hallucinations of the smell, and also with hallucinations of the taste, the patients believing they were drinking brandy instead of water.

with patients who occasionally heard voices addressing them when nobody was present.[1]

Weakness.—Great weakness, especially in the knees and hips, is a prevalent character of chronic alcoholism, which may sometimes be considered as a condition of threatened paralysis. Indeed it is occasionally so great as to interfere considerably with the various acts of voluntary motion. In one of my cases (Case 22) the patient could hardly walk, even with the assistance of a stick, and this symptom continued after he had recovered in other respects. Another (Case 13) stated that thirteen years previously he had fallen asleep drunk, and found, on awaking, that his right arm, as far up as the shoulder, was paralysed. On that occasion he had been admitted into the Westminster Hospital, and after a treatment of thirteen or fourteen weeks' duration, had completely recovered the power of the paralysed limb. From that time, however, the fingers of the right hand, with

---

[1] Magnus Huss has frequently observed a feeling in the limbs of his patients as if they were pricked with needles (formication), and also they sometimes experienced a peculiar sensation as if something was creeping along their skin.

the exception of the thumb, were drawn outward, as if having been affected with gout. A professional acquaintance told me of a remarkable instance where a soldier, having been reprimanded by an officer before his regiment on account of intemperate habits, proceeded immediately to drown his shame in beer, which he drank to an enormous extent. Paralysis of the legs and arms followed this excess, from which, however, he recovered in the course of time.

Difficulty of breathing.—Difficulty of breathing, perceived in the throat as a sensation of choking, is a frequent symptom of chronic alcoholic intoxication, entirely independent of any affection of the lungs. This symptom was particularly marked in the case of W. B——(Case 1). He described the feeling as one of choking, his breathing being quite natural for a few minutes, and then becoming suddenly checked. He pointed to the larynx as the spot where he felt an obstacle to the respiration. Upon throwing his head backwards he emitted through the mouth a quantity of air, and was afterwards able to breathe freely until the return of another spasm. This symptom is possibly owing to the frequent contact of alcoholic drinks with the glottis and epiglottis, and to the inflammation and tumefaction thereby induced; the action being somewhat similar, although in a greatly mitigated form, to that of strong mineral acids, which have been known to produce death in children from their contact with the glottis and epiglottis.[1] It may be also that these spasms result from the irritation produced by alcohol on the pharynx, inducing a contraction of the larynx by nervous reflex action. Bouchardat and Sandras have ascertained that, when alcohol is introduced into the system in excess, the blood of the arteries presents the aspect of venous blood, showing that it has not undergone the proper oxygenating process,[2] and it is possible that this circumstance should induce spasmodic contractions of the glottis through the medium of the brain and spinal chord. It appears that the

[1] 'A. S. Taylor on Poisons,' 2d edit., p. 246.

[2] On the Digestion of Alcoholic Fluids, and their office in nutrition. 'British and Foreign Medico-Chirurgical Review,' vol. ii, 1847.

habit of swallowing air, so frequently met with in those who drink to excess, is in some way or other connected with the difficulty of breathing, which in fact instead of relieving, as it apparently does at the time, it increases. In several cases, having warned patients of the importance of refraining from swallowing air, the compliance was attended with a considerable decrease of the sensation of choking. The habit of swallowing air, which after a time becomes an unconscious act, is not only extremely unpleasant from the frequent breaking of wind which it occasions, but also the source of much intestinal uneasiness, and it is necessary to stop it as soon as possible wherever it is found to exist.

CAUSES PREDISPOSING TO CHRONIC ALCOHOLISM.

It is a well-known fact that persons addicted to excessive drinking are not equally affected by it. The nature of the beverage, the quantity taken, and the time during which the habit has been indulged; the age, sex, temperament, habits, and occupation of the patient; the quality and quantity of the food taken, are so many circumstances modifying the action of alcohol on the body, and which may predispose it to suffer from chronic alcoholism.

Quality taken.—It might be considered at first sight that the more spirit or alcohol the beverage contains, the greater its deleterious action on the nervous system; and that we might consequently, from the known proportion of alcohol in fermented drinks, establish a scale showing precisely their comparative tendency to bring on chronic alcoholism. Generally speaking this rule holds good, and it will be found that raw spirits are the most hurtful; then follow wine, beer, and cider. Except, however, in a general point of view, the injurious properties of alcoholic liquors do not always depend on the proportion of spirits they contain. Macnish furnishes us with some valuable information on the effects of the different qualities of alcoholic stimulants. In his opinion, the safest way to use spirits is in the form of grog; cold toddy ranks next in safety, then warm toddy, cold punch, and raw spirits. He adds, with respect to malt liquor, "it is better to drink porter than strong ale, and advisable, when accustomed to malt liquor, to take exercise in order to avoid becoming fat and stupid, and predisposed to apoplexy." For wine drinkers, the safest wines are, according to the same author, "those possessed of the most diuretic properties, and which create least headache and fever; as Hock, Claret, Burgundy, Bucellas, Rhenish, and Hermitage. Port, Sherry, or Madeira, and sweet wines, are apt to produce acid on weak stomachs. Claret is the most wholesome wine that is known." Macnish also recommends not to drink of too many different kinds of wines at one sitting. The fact of intoxication being rapidly produced by mixtures of different wines is generally known, although usually disregarded when the temptation of a great variety of wines is offered to the guests at a dinner party. Hotel and tavern-keepers, being frequently called upon to share the drink of their customers, are very liable to suffer from the variety of the beverages they consume.

have met with cases where individuals not only indulged in many different kinds of stimulants, but mixed one liquor with another in the same glass; ale and gin, for example, being drunk together.

Cider, and such wines as possess little spirit, and are more or less acid, are frequently, and with comparative safety, used in warm countries, and, according to Roesch, may even prevent certain diseases, especially those resulting from a deficiency of the biliary secretion. Lehmann observes: "It is owing to 'cider's' great cheapness that in several eastern cantons of Switzerland, such as Thurgovie, Appenzell, St. Gall, and Zurich, the results from the abuse of alcohol, so common in other cantons, are unknown. By consulting, as I have done, the medical practitioners in this country, one becomes convinced that cider is not attended with unfavorable effects unless it be made from green fruit, or be ill prepared, or undergoing decomposition." There is no doubt, however, adds Roesch, that the abuse of cider may become the source of disease. Delbeck states, that cider does not agree with many people, producing frequently diarrhœa and various forms of indigestion.[1]

We may observe that even wines of the same quality (bearing the same generic name), and grown in adjoining districts, do not act with equal power upon the brain. Thus I am informed by a gentleman, who has had ample opportunities of making himself acquainted with the properties of Rhine wine, that the grapes grown in some adjoining districts along the Rhine do not produce wines equally heady. I believe many individuals will find that wine or beer does not exert the same action on their brain, as an equal quantity of a mixture of spirits and water, prepared so as to contain a proportion of alcohol similar to that which exists in the fermented juice of grape or in malt liquor. This circumstance may be accounted for, by assuming that the alcohol of spirits, which is distilled, differs as to its influence on the brain, from that of beer or wine which is not distilled. At all events we know

[1] Thèse sur l'influence des boissons alcooliques sur la santé, 1854.

positively that these two kinds of alcohol have not the same influence upon the sense of taste, for a wine connoisseur will be able to tell without difficulty whether distilled alcohol has been added to a sample of wine, or whether the wine contains none but its natural alcohol; this fact showing that there is certainly a difference between the alcohol of distilled and that of non-distilled spirituous stimulants. It may be also that they are not with equal readiness removed from the body or decomposed therein. The nature of the peculiar condition of the alcohol in wines, which is altered by the process of distillation, appears to be owing to an action of the other constituents of wine upon its alcohol; for it is well known that distilled alcohol, added within certain proportions to Port wine, is converted after a lapse of some years, into the non-distilled kind, its presence being no longer discernible by the taste.

Quantity taken.—I had an opportunity not long ago of overhearing a conversation between two labourers, the eldest, who appeared to have nearly reached the age of seventy, endeavouring to convince the other that the best way to attain old age was to drink freely of beer and spirits, and this, he observed, was the result of his own experience. It is certain there are exceptions to the general rule that frequent excesses in alcoholic beverages will ultimately destroy health; these exceptions, however, are but very few. We must remember that the word *excess*, in cases of intemperance, is not *absolute*, but to a certain extent *relative:* thus, an allowance of a spirituous beverage all but harmless in one instance, might be attended with dangerous effects in another. The fact that some can indulge very freely in alcoholic drinks without suffering from them, is illustrated by the following instance:—J. M'B——, aged 49; a hawker.— Was engaged thirty years ago working in a distillery at Bristol, and during the seven years he was thus employed used to drink rather less than a pint of gin daily. He travelled afterwards in Devonshire, where he drank a great deal of cider, from which he felt no further inconvenience than pain in the stomach and purging. For the last four months he has been drinking seven or eight glasses of rum a day. He exhibits no symptoms whatever of alcoholism, and applies to be treated merely for an eruption of acne on the face.

It will be observed that most of my patients suffering from chronic alcoholism, drank to a considerable extent both malt liquor and spirits. Thus W. B— (Case 1) drank one pint of gin, and two or three, and occasionally six or eight pints of beer daily. E. C— (Case 15) took half a pint of brandy and five or six pints of stout daily. G. R— (Case 18) drank from three glasses to a pint of spirits daily, and four or five pints of ale. The largest amount of alcohol I have known to be taken occurs in the cases of H. H— (Case 46), and T. D— (Case 47). The former, aged 42, contracted the habit of drinking when fourteen or fifteen years old, and from that time to the last three or four months has taken three or four pints of rum and as much beer daily.[1] The latter, also aged 42, has taken,

---

[1] This enormous quantity must be considered as an approximation to the truth. I cannot help thinking that the patient drank frequently less than his own statement might lead us to believe.

for a period of eight or ten years, an allowance of ten small glasses of brandy, as much gin, and ten glasses of beer daily. In some cases a much smaller allowance of stimulants sufficed to bring on a disordered state of the nervous system, as, for example, in that of G. J—— (Case 14), who took three or four pints of beer daily, and no spirits, and this man became intoxicated if he drank no more than two pints of ale at one sitting. The following case shows how remarkably liable some individuals are to suffer from a very small amount of alcoholic beverage. G. B——, aged 28, (Case 35), a stoker in the House of Parliament.—Admitted as out-patient at the Westminster Hospital, on Feb. 24th, 1859. Has always been of sober habits, and was only drunk once in his life, when no more than twelve years of age. His daily allowance of beer has been one pint, and he has taken no spirits. Three years ago he became a teetotaller, because he found that even so little as one pint of beer daily did not agree with his health. He has suffered from the usual symptoms of chronic alcoholism for the last three years.

Time the habit has been indulged.—A very remarkable circumstance connected with chronic alcoholism is the fact of its occurring, or of the symptoms becoming frequently aggravated, long after the habit of drinking to excess has been given up, and even, in many cases, after a complete abstinence for some time from alcoholic stimulants. As instances: C. A— (Case 2) turned teetotaller seven weeks before applying for relief at the Westminster Hospital. G. R— (Case 18) drank from youth from three glasses of spirits to a pint, and four or five pints of ale daily, until about eleven months previous to admission, when he gave up completely both beer and spirits. C. P— (Case 21) took twelve pints of beer daily for nine years, but during the last four years drank only two or three pints of ale a day. Many other instances of the same kind will be found recorded in the table. Yet, notwithstanding their abstinence, these individuals fell a prey to past excesses.

I have been led to observe that the injurious effects of the long-continued abuse of alcoholic stimulants, are frequently not developed to any extent until the occurrence of another circumstance, which is the immediate cause of the attack. It has not been possible for me to determine satisfactorily whether an attack of chronic alcoholism may supervene long after the individual has given up drinking, and without his having at all suffered from the nervous derangement known to result from frequent excesses; but this much may be safely stated: that in the great majority, if not in every case, the patient's constitution has been so far affected that the slightest cause will be sufficient to startle or frighten him, produce giddiness, headache, and keep him from sleeping at night, yet without preventing him from attending to his occupations, or proving of any material inconvenience; and such patients are very liable to a regular attack of chronic alcoholism from some cause independent of drink.

Age.—According to Macnish, a child nursed by a drunken nurse is hardly ever healthy; it is especially subject to derangements of the digestive organs, and to convulsive affections. Dr. North has observed this latter disease to be instantly arrested by transferring the child to a sober woman. Macnish adds that the habit, in some parts of Scotland, of giving raw whisky to babies a few days old, turns them pallid, and they become emaciated, fretful, subject to convulsions and every variety of disorders of the stomach, including vomiting and diarrhœa, which may end in death. The following investigation, undertaken by Dr. Hunter, reported by Macnish and by Carpenter, shows that alcoholic drinks, even in moderate quantities, do not agree with young children. He submitted to experiment his two children, both of them having been previously unused to wine. To the one, a child of five years old, he gave every day a full glass of sherry; to the other, of nearly the same age, he gave an orange. In the course of a week a very marked difference was perceptible in the pulse, urine, and evacuations from the bowels of the two children. The pulse of the first child was raised, the urine high-coloured, and the evacuations destitute of their usual quantity of bile. In the other child no change whatever was produced. He then reversed the experiment, giving to the first the orange, and to the second the wine, and the result corresponded; the child who had an orange continued well, and the system of the other became straightway disordered, as in the first experiment.

A young man betaking himself to the disgusting habit of drinking to excess may be considered as very liable to become intoxicated. Should his constitution not be strong and healthy, or should he not be accustomed to active habits, the first occurrence of disease may prove fatal, or, at all events, give him such warning as will effectually prevent his returning to drink. But there are other young drunkards, gifted with strong and healthy constitutions and engaged in occupations requiring great muscular exertion in the open air, who are enabled to rid themselves rapidly of the alcoholic poison; for instance, men employed as coal porters, hawkers, labourers, will be able to resist for many years the baneful consequences of intemperance; but at forty-one years of age, those who have habitually indulged to excess in alcoholic liquors begin to suffer, probably because that age is the time of life when youth and health often begin to depart. One individual becomes a prey to gout, another to rheumatism, another to bronchitis, and another to disorders of digestion. Now, those very diseases are known to predispose greatly to chronic alcoholism, and it is therefore not to be wondered at that this disorder should usually make its appearance at the above-mentioned period. The youngest patient coming under my treatment for chronic alcoholism was eighteen years of age, and the oldest, seventy-five.

Sex.—Women appear to be much less subject to suffer from the long-continued abuse of alcoholic liquors than men; indeed, I have only seen one case of chronic alcoholism among the female sex, which, however, I had not an opportunity of attending. Magnus Huss accounts for his having met with but few cases of chronic alcoholism among females, by assuming that the male sex is not more predisposed to alcoholism than women, but that men indulge in alcoholic liquors more than women.

Of 139 patients treated by Huss, during three years, there were 123 men and 16 women. It is well known that delirium tremens is not frequent among females; Roesch states that, in his opinion, this phenomenon is not owing to the circumstance that fewer women drink than men, for the disproportion is too considerable not to depend upon other causes. In 170 cases of delirium tremens, observed by Rayer, there were only 7 females affected; Bang, at Copenhagen, only observed 10 among 456 patients; Hœgh-Guldberg only noticed one case in 173; Kruger Hausen, 1 in 16; and the directors of the hospital at Christiana, 1 in 11.

Temperament.—According to Delbeck, the more plethoric and sanguine the temperament, the more the nervous system is irritable, and subject to suffer from alcohol; individuals having a lymphatic temperament, being better able to stand spirituous beverages. Magnus Huss expresses a similar opinion; he classifies the different temperaments as follows, according to their degree of influence in predisposing to chronic alcoholism: 1st, the sanguine temperament; 2d, the phlegmatic; 3d, the bilious; 4th, the lymphatic; 5th, the nervous.

Habits.—The habits of an individual are among the most important of the causes predisposing to chronic alcoholism. As a general rule, habits of indolence and idleness, independently of their acting as a strong inducement to drink, favour slow poisoning by alcohol. Where the disorder is limited to the mildest symptoms, I have repeatedly observed it to be checked in a remarkable degree by having recourse to exercise of the mind and body, and I have noticed individuals accustomed to hard work become affected with symptoms owing to past excesses, because they had no longer any work to do.

Smoking tobacco.—Tobacco, and more especially that of the strongest kinds, decidedly predisposes the nervous system to suffer from the long-continued abuse of spirituous drinks, an observation which is not without importance and interest considering that the habit of drinking is frequently accompanied by that of smoking. If it be remembered that poisoning by tobacco fumes is attended with giddiness, trembling, and other symptoms referable to a disordered condition of the nervous system, it will obviously follow that the habit of smoking, and especially of smoking to excess, will act conjointly with that of drinking in bringing on an attack of chronic alcoholism. Here are a few cases illustrating this fact.

H. E— (Case 33), a clerk, admitted on the 21st of February, 1859, drank at Christmas last a considerable amount of ale and gin, and has been in bad health since that time; suffering from trembling in the morning, weakness, and loss of memory, and has lately been unable to sleep at night. This patient has contracted the habit of smoking a great deal of shag tobacco, to which circumstance he himself partly ascribes his illness. E. C— (Case 15), six years ago took to the habit of drinking about half a pint of brandy daily, and five or six pints of stout; after keeping up this allowance for nearly four years he reduced it to eight pints of porter daily, and continued with that amount up to the time of his admission on the 16th of December, 1853. He has suffered for the last four years from chronic alcoholism, one of the symptoms being trembling in the morning, especially when he has drank freely the evening before. He states that he has been a hard smoker, and has observed that smoking increased the trembling. A. P—— (Case 41), æt. 40, a gas-fitter, admitted on the 21st March, 1859, had been in the habit of drinking for the last two or three years six or seven pints of beer daily. Six or seven months ago he began to exhibit symptoms of chronic alcoholism, and about six weeks previous to admission, when smoking three quarters of an ounce of tobacco per day, he suddenly became worse, his nervous system having evidently been much affected by the use of the tobacco.

The habit of smoking being one so prevalent among those who are fond of alcoholic stimulants, and its effects being so directly connected with the slow poisoning from alcoholic beverages, an inquiry into this subject will not be out of place on the present occasion.

It is difficult to conceive why boys take so much pleasure in smoking cigars or a pipe, when out of sight of their parents or guardians; but vanity and pride probably here act a prominent part. Children desire to imitate their elders, and show each other they can colour a pipe, or smoke a cigar to the very end, even if they have to pay the penalty of nausea and vomiting. Thus, the individual contracts a most pernicious habit, which he will find extreme difficulty in breaking off if obliged to do so in after life. When smoking is becoming a habit, it gradually ceases to cause sickness, although still occasioning nausea and giddiness, and finally the nausea and giddiness disappear, returning only on special occasions. It is then that tobacco smoking produces pleasurable sensations like those resulting from the narcotism of opium, when the imagination changes each successive puff of smoke into every description of fanciful objects, when the bachelor forgets his solitude, the mind its troubles, and the body its pains. A painful impression on the mind is certainly soothed by smoking, and this is a frequent cause of the habit being contracted. One man will take to smoking to drown the disappointment of unsuccessful labours, and another to allay the affliction from the loss of a friend.[1] Smoking is also a frequent habit among those who are called upon to exercise much mental exertion, because it appears to possess the power of resting the mind when tired. When the body and mind are excited, as is usually the case after dinner, or *post pocula*, or in a convivial meeting of friends, smoking is often resorted to as an instinctive means of keeping the excitement within certain limits. The very prevalent habit of smoking after dinner must have for its principal object that of allaying the discomfort arising from the stimulating action of the meal. At dinner a glass of Sherry or Madeira immediately follows the soup, these, along with Port, being our strongest wines. At first sight, it does not appear rational to commence with the strongest wine, which must necessarily impair more or less the taste and stimulating effects of the weaker, such as Hock and Claret, that

---

[1] Smoking appears to have the property of diminishing the power of mental abstraction; it is probably on this account that when the mind is haunted by some painful idea, the act of smoking assists the effort of the will to shake it off.

may follow. But it is found agreeable to begin by assisting the appetite with a powerful stimulant, and the more alcoholic the beverage, the better it answers the purpose, so much so that some are not satisfied with Sherry after the soup, but begin dinner with a glass of spirits. The stomach is thus induced to take more than is required, and after dinner a sensation of fulness is felt, which is conveniently relieved by a cup of strong coffee; but now an uneasy sensation of heat and excitement is experienced, from which the body partly recovers by means of a full-flavoured cigar or a pipe.

Finally, it is remarkable how much certain individuals can smoke on special occasions without its producing giddiness or sickness, or any unpleasant feeling, the same persons being in general easily affected by tobacco smoke. In all cases where the body and mind are excited within certain limits, smoking will be most easily tolerated; as, for example, after dinner during the excitement from eating and drinking, in a party attended with the enjoyment of conversation, or a sporting excursion when under the influence of sport and exercise and the stimulating power of the open air.

I may perhaps be allowed to conclude these observations by advising young men, whatever may be their position in life, not to take to smoking, as such habit is certainly not conducive to health. To adults of a sound constitution, who have contracted the habit of smoking to a moderate extent, and do not feel the worse for it, I would recommend to smoke only after a meal, and that but seldom.

Profession.—Such professions as exhaust the body, from their requiring great muscular exercise, especially under circumstances which excite excessive perspiration, favour the early development of chronic alcoholism; and occasionally in these instances a very small proportion of alcoholic beverage will be hurtful. Thus, individuals obliged to remain long before a blazing fire, as furnace-men and engineers on board steamboats or in factories, are very liable to suffer from spirituous stimulants. For example, the following cases may be mentioned: G. B— (Case 35), a stoker in the House of

Parliament, who, at the early age of twenty-three, laboured under chronic alcoholic intoxication, although he was in the habit of taking no more than a pint of ale daily, and no spirits, and had become, moreover, a teetotaller three years previous to his admission into the hospital. T. S— (Case 3), an engineer on board a steamboat, consulted me for chronic alcoholism at the age of thirty-three. He had suffered previously from several attacks of delirium tremens. I have already recorded the case of an engine-driver contracting chronic alcoholism at fifty-six, his immunity arising probably from a very robust state of health. The average age of these three cases is thirty-seven, while that of the forty-seven patients whose ages have been reported in the synoptical table is forty-one. Professions affording very little or no exercise of the mind and body predispose also to suffer early, and consequently when young, from the habit of drinking. Thus, clerks, tailors, and other tradesmen, cab and cart drivers, are very liable to disorders of the nervous system from intemperance, and the table appended to this work offers these illustrations, showing the truth of the foregoing observation: E. B—— (Case 9), a tailor, age 41; G. R—— (Case 18), general dealer, age 38; G. M—— (Case 19), cushion maker, age 33; W. F—— (Case 24), a carter, age 34; J. H—— (Case 25), a shoemaker, age 27; W. P—— (Case 26), a cabman, age 29; H. E—— (Case 33), a clerk, age 24. There are, however, two instances reported of individuals of sedentary occupations suffering from chronic alcoholism at an advanced age: W. J—— (Case 5), a shoemaker, age 75, and D. B—— (Case 27), greengrocer, age 72; and if, reckoning the last two cases as exceptional, we take the average age of the patients of this class, we shall find it to be thirty-two, whilst the average age of the forty-seven patients is forty-one. Other occupations requiring much exercise in the open air also furnish our hospitals with a great number of cases of chronic alcoholic intoxication, but these individuals are not affected so young as those previously mentioned, their average age being thirty-nine. I allude especially to labourers, coal-porters, sailors, hawkers, and carriers. Finally, there are occupations and professions entailing a moderate amount of exercise, as those of butchers or policemen, which afford but few instances of the disease.

There is no doubt but that in the higher class of society a great number of cases of chronic alcoholism may be observed, although the disease is certainly most prevalent among those who attend our hospitals. Dr. Budd, alluding to the indigestion of drunkards, connected with symptoms of chronic alcoholism, observes[1]—" The kind of disorder we are considering is now seldom met with except in the lower ranks of life. Half a century ago, hard drinking was common in the upper classes, and men of fortune were often sent to Bath, to restore the tone of their stomach by drinking of its waters." It is true that intoxication is now very seldom met with in the upper classes of society; but, from the great predisposition of many to suffer from alcohol taken even in small doses, there must still exist a comparatively vast number of individuals, in comfort-

---

[1] Dr. Budd, 'On the Organic Diseases and Functional Disorders of the Stomach,' p. 290.

able as well as wealthy circumstances, whose nervous system becomes affected from the effects of the long-continued habit of drinking wine or spirits.

Circumstances connected with the food taken.—In addition to the above, there are other circumstances which appear to increase to a considerable extent the tendency of alcoholic drinks to produce chronic alcoholic intoxication; namely, drinking early in the morning before breakfast, and consequently on an empty stomach, and living on a spare solid diet; in the latter case the beverage constituting nearly the whole of the food taken.

It is a prevalent habit for labourers, in some parts of the country, to sleep in the tap-room of public-houses, where they have a free night's lodging; and it is an interesting sight to see such work-worn individuals lying about in winter near the fireside on the floor, tables, and benches, and enjoying as complete a rest as if they were in a comfortable bed. Early in the morning they awake, and most of them begin with beer or spirits; commodities within their immediate reach, and which they suppose will give them an appetite for breakfast, and keep out the cold for the whole of the day. During the week, the toils of the day call them out into the fields, and they must leave the table; but on Sunday, these same men continue imbibing the poison from morning to evening, and taking very little food; as early as 9 or 10 o'clock a.m. symptoms of drunkenness in the tap-room may be observed, and no doubt that with many labourers the morning dram on an empty stomach is the cause of their suffering, sooner or later, from chronic alcoholism. There is a class of men whose only nourishment for days in succession is beer. I allude more particularly to those who are employed in breweries. The stomach of these individuals becomes filled with nothing but beer, which must be absorbed with a much greater rapidity than during the normal process of digestion, when fluids taken as drinks become intimately mixed up with the food; and it is in no degree remarkable that this constant passage of alcohol from the stomach into the blood and brain should give rise to cerebral symptoms. Moreover, the appetite of such men for solid food is quickly blunted, not only on account of the physiological fact that alcohol diminishes the waste of the body, and consequently its requirements for new materials, but more especially from the long-continued action of the fluid on the nerves and lining membrane of the stomach. The first contact of alcoholic fluids with the mucous membrane of the stomach occasions in that organ a state of irritation and excitement, but when drinking has been continued for a length of time, the stomach becomes gradually benumbed, losing more or less of its sensibility and motility, which partly accounts for the diminished appetite and want of digestive power. The nature of the direct action of alcohol on the human stomach may be inferred from some very interesting experiments of Baron Humboldt. This eminent physiologist immersed the crural nerve of a full-grown and lively frog in alcohol. If the leg was already exhausted by galvanization the alcohol evidently increased its excitability, and this increase lasted for a time on the nerve being quickly removed from the stimulating fluid; but if the nerve was left in it for some time, its excitability became completely exhausted. The application of alcohol exhausted instantaneously the excitability of young animals,—birds, worms, and insects. If the tail of an earth-worm or leech was dipped for only four seconds in alcohol, it became stiff and inexcitable as far as it had been immersed, and, although in frogs and puppies submitted to the same experiment this state of rigidity could sometimes be removed, in the earth-worm or leech it remained permanent.[1]

In addition to this morbid state of the organs of digestion occasioning a weakness of their digestive power, the diminished appetite of those who live principally on alcoholic drinks causes a deficiency of the solid food taken. The strength of the body is also thereby impaired, thus predisposing it to suffer from chronic alcoholism.

[1] Dr. Carpenter, 'On the Use and Abuse of Alcoholic Liquors,' p. 8.

IMMEDIATE CAUSES OF AN ATTACK OF CHRONIC ALCOHOLIC INTOXICATION.

A question of great practical importance now suggests itself to our notice. *What is the nature of the last and immediate cause which determines an outbreak of the disease?*

This question is one very difficult to answer, for in many cases the disorder creeps on slowly, under the influence of predisposing causes, so that when the urgency of the symptoms obliges the sufferer to have recourse to medical advice, he cannot state precisely the time when the illness began, neither can he ascribe it to any particular cause, except that of intemperance. The usual expression of such patients is: that drinking does not agree with them; and under these circumstances many drop the habit entirely, or greatly diminish their allowance. Thus, for example: G. R— (Case 18), found his health was gradually giving way from intemperance, and gave up drinking about eleven months previous to admission. In the case of C. P—

(Case 21), no other immediate cause of the illness could be detected than hard drinking; he had fallen ill four years before admission, and during nine years had been in the habit of taking no less than twelve pints of beer daily.

An exceptional excess.—An attack of chronic alcoholism will frequently be determined in habitual tipplers by an exceptional excess and a severe fit of drunkenness. Thus, for instance: J. W— (Case 11), age 18, after drinking all day from ten in the morning to twelve at night, was suddenly seized next morning with symptoms of chronic alcoholism. G. F—,[1] admitted on the 10th of January, 1859, committed an extraordinary excess during three days—at Christmas last, when he took six or seven quarterns of raw gin, and about three pints of beer daily, after which he suffered from chronic alcoholism. He is not an habitual drunkard, and takes usually a pint of beer daily. W. P— (Case 26), two nights previous to falling ill, indulged more than usual, and was drunk.

[1] Not reported in the Table from want of completeness of notes.

B. L— (Case 36), admitted on the 28th February, 1859, was attacked with chronic alcoholism four years ago, but is worse now than previously, from drinking at Christmas last more than his usual allowance.

Coexisting disease.—An attack of chronic alcoholic intoxication is frequently brought on by another disease, whether this disease be produced directly by excessive drinking, or whether (which seldom happens) it be quite independent of intemperate habits. The same observation applies to delirium tremens, a disease very closely allied to chronic alcoholism.

We read in the work of Roesch: "An individual in the habit of drinking to excess, affected by a chronic or acute disease, which cannot be attributed directly to intemperance, often suffers from peculiar symptoms, and the alcoholic poisoning which had remained latent breaks forth suddenly. This happens in the case of delirium tremens, especially when drunkards become affected with inflammation of the lungs." In one case, Roesch has observed the disease to be brought on by very painful attacks of rheumatism.

Dr. Schmidt, at the General Hospital of Hamburg, has seldom seen a perfectly genuine case of delirium tremens; this affection appeared to him to be brought on usually by powerful emotions, and by external lesions, and inflammations, especially of the chest. He reports, that Channing states having met with delirium tremens attended six times on seven with an affection of the chest. The habit of drinking may therefore be indulged in with impunity until the occurrence of some affection of the lungs, or rheumatism, or another illness, causes the appearance of those symptoms directly referable to the abuse of alcoholic stimulants. The two following cases, where chronic alcoholism was occasioned by attacks of gout, are well calculated to illustrate the truth of the foregoing observation. T. D— (Case 47), a cab-driver, age 42, admitted on the 25th of April, 1857. He began drinking five or six pints of ale daily, which he kept up for upwards of five years. Since that time he increased his allowance to ten small glasses of brandy, as much gin, and ten glasses of ale daily. He went on taking this large quantity of ale and spirits for eight or ten years, being none the worse for his intemperance, until he was seized with gout two years before admission. The disease lasted a fortnight, and was followed by a severe attack of chronic alcoholism, from which he has suffered ever since.

D. B— (Case 27), a green-grocer, age 72, admitted as out-patient on January the 17th, 1859. Took to the habit of drinking to excess when a young man, his average allowance being four or five glasses of spirits and three or four pints of beer daily; but for the last fourteen or fifteen years he has given up hard drinking, and reduced the amount to a pint of beer daily and an occasional glass of spirits. On rising in the morning he trembles a great deal for rather less than half an hour; perceives occasionally imaginary objects, and a mist forms at times before his eyes; hears absent people addressing him; does not complain of headache or giddiness; has very little sleep—according to his own expression, night after night he lies awake in the greatest agony. Frequently suffers from sickness and vomiting. These various symptoms have assumed an intermittent form, lasting for a month, and occurring at intervals of about a week. The patient suffers, moreover, from gout, as will be seen from the following additional statement. There is considerable pain in the ankles, knees, and hands; the thumb of the right hand is now greatly thickened; he feels an acute pain in the great toes of both feet, and can only walk with the assistance of a stick. These symptoms invariably occur with those of chronic alcoholism. As soon as the attack of gout passes off, the patient becomes also free from the other disease, and as often as the gout returns the nervous system again becomes affected. In this case, therefore, the occurrence of gout was obviously the immediate cause of the attack of chronic alcoholism. I have had opportunities of observing a similar effect resulting from rheumatism, affections of the lungs, and gastritis.

Independently of the fact that the poisonous action of alcohol is often developed by the occurrence of another disease, chronic alcoholism, it may be observed, is very frequently accompanied by a secondary disorder, which it is difficult to connect with the outbreak of the former; and, moreover, the morbid condition of the nervous system, owing to long-continued intemperance, is not unfrequently concealed by the coexistence of another disorder, exhibiting symptoms of a more prominent and decided character; I need not insist on the importance of making in these cases a correct and complete diagnosis. A glance at the synoptical table will show the tendency of bronchitis to accompany chronic alcoholism. For instance, J. H— (Case 20), suffering from chronic alcoholism, is attacked every winter with bronchitis; J. R— (Case 23), admitted at first for bronchitis, to which he has been subject for several years; but it was found subsequently, he also laboured under chronic alcoholic intoxication; J. H— (Case 25), was treated at first for bronchitis, shortly after his admission it was discovered he was suffering from chronic alcoholism.

Affections of the stomach are so very frequently the result of excessive drinking, that it is difficult to determine whether they are actually the cause which determines the outbreak of chronic alcoholism or constitute merely a coexisting disease. Dr. Budd gives us the following correct and interesting account of the disorder of the stomach so frequently met with in drunkards : " The chief characters of this (indigestion of drunkards) are want of appetite, and vomiting or dry retching in the morning, with a white or furred tongue, and a slow pulse; the power of digestion is much enfeebled, and if the patient eat at any time, what for others would be a very moderate meal, he is apt to vomit soon afterwards, and to be troubled with pain and flatulence.

" This disorder, like the vice from which it springs, is most frequent in men of middle age, and is generally associated with more or less of that strange and peculiar disturbance of the nervous system which hard drinking brings on, and of which the most striking effects are inability to sleep or sleep broken by frightful dreams, despondency in the morning, and tremulousness of the hands and tongue."

I have often asked my patients whether the loss of appetite, pain in the stomach, and vomiting preceded the appearance of the nervous symptoms, but have very seldom obtained a perfectly precise and satisfactory answer; opportunities have, however, offered for my observing, that an increase of the morbid condition of the digestive organs may be attended with a marked progress of the disturbance of the nervous system, and that an improvement of the digestion may be accompanied with a diminution of the nervous symptoms. A glance at the accompanying table will show the frequency of disordered digestion in cases of chronic alcoholic intoxication. J. T— (Case 13) complains of pain in the stomach as soon as he has ate or drank. L. M— (Case 17), on falling ill, first complained of pain in the stomach after taking food, and restlessness at night. W. F— (Case 24) suffers from pain in the stomach and sickness, increased after taking food. W. P— (Case 26) complains of pain in the stomach, increased by eating. D. W— (Case 28) has lost his appetite and feels sick in the morning. B. L— (Case 36), appetite deficient, great pain in the stomach after eating.

Cerebral concussion.— I have met with an instance of chronic alcoholism suddenly occasioned by a fall, producing at the time cerebral concussion. This case was that of A. P— (Case 41), a gas-fitter, in the habit of drinking six or seven pints of beer daily for the last two or three years. Six or seven months previous to admission, being then in perfect health, he fell from a height of twenty-five feet. The fall must have been attended with cerebral concussion, as he stated that "his nerves appeared to have been shaken." From that time he began to suffer from symptoms of chronic alcoholism, which were subsequently increased by smoking. On admission he complains of want of sleep, with wandering of the mind and giddiness; trembles and feels sick in the morning, has no appetite, and suffers from pain in the stomach. I beg to remind the reader that the symptoms of cerebral concussion and of drunkenness are very much alike, and if it be admitted that a fit of intoxication is frequently the immediate cause of an attack of chronic alcoholism, a similar result may be expected from cerebral concussion. The analogy between the symptoms of drunkenness and cerebral concussion has been observed by Sir Benjamin Brodie. (See 'Phil. Transact.' for 1811.) He states, p. 181: "Concussion of the brain, which may be considered as the slightest degree of injury, occasions a state of mind resembling intoxication, and the resemblance in some instances is so complete, that the most accurate observer cannot form a diagnosis except from the history of the case."

It may be stated, I believe, as a general rule, that any disease affecting subjects given to drinking, will, by weakening the body, more fully expose the nervous system to suffer from intemperate habits.

AFFECTIONS OF THE NERVOUS SYSTEM RESEMBLING CHRONIC ALCOHOLIC INTOXICATION.

There exist certain disorders of the functions of the nervous system owing to other causes than intemperance, and closely resembling chronic alcoholism; these are produced by long-continued and excessive intellectual exertions and by sudden and violent emotions. These affections exhibit symptoms which bear a close analogy to those of chronic alcoholism, and I have treated them successfully by the same means.

I have preserved notes relating to several patients who suffered from disorders of the nervous system owing to excessive mental exertions, the principal symptoms being headache and giddiness, with much nervous uneasiness, and sometimes hallucinations; want of sleep, and occasional palpitations.

In November, 1854, I attended Mrs. B——, aged 24; for the last three months she had been greatly engaged with literary pursuits, and ascribed her illness to excessive reading and exertion of the mind; she suffers from headache, giddiness, and great nervous uneasiness; she cannot sleep at night, her appetite is indifferent, and she is subject now and then to palpitations.

The following is a remarkable case which came under my notice, where the disturbance of the nervous system, from excessive mental exertion and anxiety, had reached such a pitch as to become actual mental alienation.

C. T— aged 41, a gas-rate collector, apparently in easy circumstances, of a very nervous temperament and excitable temper. He had also been a gas inspector and subject to much anxiety connected with his business. There exists a slight tendency to suicide, but he observes that he never could or would destroy himself. He is accompanied by his wife, who considers all he says as perfect truth. C. T. expresses himself fluently, and his memory appears very lucid, from the minute account he gives of his case. On September 13th, 1855, he was engaged the whole day in writing, and appeared much excited; he went to bed at 11 o'clock, and the next morning rose at 3. At 9 a.m. he proceeded to his business, leaving his papers in great disorder, which he had never done before. He returned home at 5 p.m., and asked for some tea. Having shortly afterwards kissed his children, (his wife was then absent), he left home, though for no definite purpose, taking a few pounds with him. He wandered about all night, and found himself in the morning at Hampton Court. He continued walking, not knowing where he was going or why he had left his house, and after some days arrived at Southampton. He next embarked on a packet leaving for France, and resumed his rambles in Normandy, where he suffered great privations, sleeping in the fields and on the high road, and living upon nothing but turnips, apples, and water. Having remained abroad for about one month, at the commencement of November he returned to Plymouth, stopping on his way at Jersey; and then wandered to Teignmouth, Totness, Exeter, and from thence to Bristol and Clifton. He still had some money with him, although he thought he had none, and that he wanted none. With the view of joining his family on the birthday of one of his children, this unfortunate man returned home, and arrived late at night at his own door; the person who opened it unluckily did not recognise him, and closed the door upon him. He then set off again on his rambling life, and walked to Brighton with the object of visiting some relations who he knew would be kind to him; but when at Brighton he never called upon them, although he passed under their very windows; he does not know why he did not see them. From Brighton he went to Shoreham, to Worthing, and to Preston, where, his money being entirely spent and having no food, he began to suffer the most bitter pangs of hunger. He used his utmost endeavours to obtain food from charitable institutions, and was directed to Arundel, whence, finding no relief, he proceeded to Chichester. In this town the sufferings from hunger were such that he was induced to procure food by dishonest means; he was taken up and imprisoned, with hard labour, for twelve days. When liberated he immediately started for London, to which he walked in two days, the distance being sixty-four miles. During the journey he took nothing but bread and water; he proceeded immediately to his house, and arrived there on the 27th of November, having been absent from home more than two months.

It was not until this patient had been thoroughly cross-questioned that I could believe the foregoing account; in order to obtain a corroborative evidence, I examined his naked feet, and found the soles hardened, resembling tough leather, and showing that he had certainly walked a considerable distance.

Sudden grief may produce symptoms identical with those of chronic alcoholism; indeed, in the following case I was with some difficulty convinced that the patient was not addicted to the habit of excessive drinking.

R. M.——, a coal-porter, admitted as out-patient on March 10th, 1859, is of regular habits, and has never been addicted to drinking; has not been more than twice in his life the worse for liquor. Lost his wife eight months ago, at which he was greatly distressed, and since then has suffered from the frequent occurrence of giddiness, headache, and black specks before his eyes. He trembles a great deal, especially in the morning; cannot sleep well at night, but his sleep is not disturbed by nightmare. Is very weak, yet has not given up work. Has smoked a great deal after the death of his wife, for the purpose, he says, of driving care away; he now perceives that smoking makes him very giddy, and he is gradually giving up the habit.

There are other diseases having symptoms in common with chronic alcoholism, such as chorea, hysteria, the sequel of an attack of apoplexy, and tremors from poisoning with lead or mercury; but it would be inconsistent with our subject to enter upon any observations on these affections.

Prognosis.—As may be anticipated, the successful treatment of chronic alcoholism depends, first of all, on the habit of drinking to excess being given up. As soon as this result is obtained, or if the patient applies for relief after having, of his own accord, ceased drinking, a favorable issue of the disease is to be expected. In those instances where the complaint is unattended with other disorders, and if the patient be not in reduced circumstances, the treatment is very successful, the sufferer being cured of every symptom; the weakness of the limbs, so prevalent in this disease, is sometimes, however, troublesome to overcome entirely. But in cases where the disordered nervous system, consequent on intemperance, is accompanied by another disease, the treatment may have to be continued for some length of time, and will sometimes leave the patient still labouring under the secondary affection. A remarkable instance of this kind, which has already been alluded to, was that of an individual subject to gout, and who was seized with chronic alcoholism at every successive return of the gouty attacks. He was treated for chronic alcoholism, and gradually the symptoms of the morbid condition of the nervous system disappeared, although the attacks of gout returned. Disease of the liver appeared in one instance to keep up chronic alcoholism, although in the course of the treatment a marked improvement of the nervous symptoms was obtained. Inflammations of the lungs also increase the difficulty of arresting and curing the disease. The same observation applies to disorders of the digestive organs, and rheumatic affections. But it must be understood that an improvement in the secondary disease is usually attended with a relief of chronic alcoholism.

TREATMENT OF CHRONIC ALCOHOLIC INTOXICATION.

The treatment of a chronic disease is usually attended with much difficulty, and a long period will generally be required to effect a complete cure, even when such an end is attainable. I have succeeded, however, by a simple method, and within a comparatively limited period, in restoring health in cases of chronic alcoholism of the severest description. The treatment is to be considered under two heads: 1, the patient must be induced to give up the habit of drinking, unless he has done so previously; and 2, the inordinate desire for alcoholic stimulants having been subdued, the next point is to arrest the disease. It is usually considered that the habitual abuse of spirituous liquors becomes so inveterate as to defy all control; and the following case, related by Macnish, bears upon the subject: "A gentleman accustomed to drink to excess answered to the exhortations of a friend—'If a bottle of brandy stood at one hand, and the pit of hell yawned at the other, and if I were convinced I should be pushed in as surely as I took one glass, I could not refrain.'" I have observed, however, that when health is evidently giving way from over-indulgence in spirituous beverages, drinking is often given up spontaneously, or, at all events, considerably diminished. By consulting the analytical table the reader may notice that in the cases of W. H— (Case 7), J. T— (Case 13), E. C— (Case 15), L. M— (Case 17), G. R— (Case 18), C. P— (Case 21), J. H— (Case 30), and others, the allowance of spirituous beverages had been spontaneously diminished, or altogether stopped, long before applying for medical advice. In these instances, the patients stated they had given up excessive drinking because they considered their health was suffering from it. When the habit of drinking has become inveterate, I have invariably found it of great importance to begin by obtaining the full confidence of the patient. Dr. Trotter, describing the means of checking this pernicious habit, observes —" When the physician has once gained the full confidence of his patient, he will find little difficulty in beginning his plan of cure. . . . This confidence may sometimes be employed to great advantage when your regimen is in danger of being transgressed, for frequent relapses and promises repeatedly broken, will, in such situations, render the physician's visits a work of great trial to his patience. This disease (I mean the habit of drunkenness) is like any other mental derangement — there is an ascendancy to be gained over the person committed to our care, which, when accomplished, brings him entirely under our control."

It will be necessary to begin by impressing upon the patient's mind that he is really suffering from the habit of drinking too much, and that it is of no use to commence a medical treatment unless he be decided to abstain from spirituous stimulants, or, at all events, diminish considerably his usual allowance. If he appear ready to act according to this advice, it will be desirable to inquire into the cause of the abuse, and remove it, if possible. Should an individual drink from habits of indolence, let him exercise his mind and body; if another drinks to drown dull care, let him have amusements; if intemperance results from frequenting bad company, let such acquaintances be dropped; if spirituous liquors be taken as a preservative against cold, let the clothing be increased, and a more nutritious solid diet taken.

But, as Roesch observes,—"The inordinate desire (passion) for drink becomes itself a cause of drunkenness, for when the body has accustomed itself to spirituous liquors, it can no longer do without them." In such cases little can be derived from an attempt to stop the cause of intemperance, and other means must be adopted. Macnish relates the case of an individual in Maryland, much addicted to drinking, who heard one evening a noise in his kitchen, and, on opening the door to ascertain its cause, found his servants laughing at the exhibition of a young negro, who was engaged in mimicking his master under the effects of liquor; this comic scene produced such an impression on him that he was never known to be drunk again. Dr. Pitcairn is reported to have cured a Highland chieftain by obtaining from him a promise that he would drop a little sealing-wax every day into his whisky-glass, when after a time, the sealing-wax having filled the glass, he could drink no more. A gentleman, to wean himself from drinking, used to add a glass of water to his bottle for every glass of Hollands he took out, till at last the drink got so insipid that he could no longer go on with it. An American physician, Dr. Kairns, taking advantage of the nauseating properties of tartar emetic, has advised this drug to be mixed with the patient's drink; but this process is condemned by Roesch, who observes that the trick will soon be found out, and drink procured that has not been previously medicated. Tartar emetic has, besides, the disadvantage of acting injuriously on the patient's health.

There is in every class of society a number of persons who, although they do not become intoxicated, suffer from chronic alcoholism, from drinking more spirits, wine, or beer than agrees with their health. Most of these persons lead a useful and active life, and apply for medical advice, being quite unaware of the cause of their illness. Many of the upper ranks of society are thus seized with symptoms of chronic alcoholism. The habit of indulging freely in wine at frequent dinner-parties, of drinking wine at lunch, of taking occasionally a glass of wine between meals, or of sipping every evening two or three glasses of sherry and water, or brandy and water; the usual good living at the officers' mess or at the clubs; the custom which exists for commercial travellers, not only of using freely stimulants at dinner, but also of drinking wine with their customers when transacting business, and finding, of course, an equal pleasure in these potations—all these various circumstances, and many others besides, are quite sufficient to bring on an attack of chronic alcoholism when an individual is predisposed to the disease. Drinking is not usually in these cases an indomitable habit, and accordingly, the patient will gladly give it up if he feel certain that by so doing his health can be improved.

A general opinion is very prevalent, that an individual, whose health suffers from the habit of drinking to excess, may cure himself by taking to sober habits or giving up drink entirely; and, it is with this end in view, that many are induced to join temperance societies. I have shown, however, the fallacy of this idea, not only by reporting Dr. Carpenter's opinion on the subject, but also by bringing forward a certain number of instances where patients applied for advice subsequently to their having partly or entirely given up the habit of hard drinking. It is consequently necessary, after putting an end to the habit of drinking to excess, to adopt an active medical treatment.

The nature of the diet to be recommended will vary according to circumstances. In general, those suffering from chronic alcoholism have little or no appetite. The first step in such cases is to begin by relieving this symptom, and then recommend a nutritious and easily digestible food, such as broth, sweetbread, carefully roasted meat, provided it be lean, on account of the secretion of bile and pancreatic juice being probably much below their usual standard. Strict attention must be paid to this part of the treatment, as an insufficient and unhealthy kind of food undoubtedly predisposes the body to suffer from the morbid effects of the long-continued habit of indulging to excess in spirituous beverages; and, on the contrary, a healthy and nutritious diet is known to be beneficial.

The following illustration shows the importance of attending to diet in cases of chronic alcoholism: J. T— (Case 13), a poor labourer, formerly addicted to drinking to excess, had been out of work for a long time, and could procure little besides tea and bread and butter; he complained of the usual nervous symptoms of chronic alcoholism, and also of want of appetite and great pain in the stomach. His circumstances, however, gradually improving, he was able to afford better food, when the appetite became keener, with a simultaneous diminution of the pain and of the nervous symptoms. Not long afterwards, being again obliged to reduce his diet, the pain, want of appetite and symptoms of chronic alcoholism returned.

With respect to the use of alcoholic stimulants, if the patient has completely given them up for some time, and entirely lost his taste for liquor, I have been in the habit of recommending about a pint of bitter beer daily, to be taken at meals; but in so doing much care is required, as some patients, who formerly could drink hard without being the worse for it, become liable to be easily affected by alcoholic beverages, even of the mildest description. The reader need hardly be reminded that tea and coffee are excellent substitutes for alcoholic drinks, which they resemble, not only by their stimulating powers, but also by their remarkable property of diminishing the waste of the body, thus effecting an indirect process of nutrition. In certain countries, such as Norway, where in many districts alcoholic drinks are seldom to be obtained, I have observed coffee and milk to be extensively used, both as food and in the place of beer and spirits. But although tea or coffee may be resorted to as beverages, they cannot be introduced at dinner instead of wine and beer. Many do not like soda water; toast and water is unpalatable; such beverages as lemonade are unpleasant when taken with solid food, and no choice therefore seems to remain but to drink pure water, which, although insipid at first when taken by those who are in the habit of drinking beer and wine, soon becomes agreeable and refreshing, especially if ice cold.

We shall now proceed to inquire into the therapeutical treatment of the morbid condition of the body described in the preceding pages.

Chronic alcoholism is not to be cured, like delirium tremens, in the course of a few days; for although under an appropriate treatment a marked improvement may in most cases occur after a short time, a much longer period will be required to restore the patient to perfect health.

Magnus Huss has derived very satisfactory results from the treatment of chronic alcoholism with fuseloil (fuselöl, or Fermentoleum solani), given in the form of pill with Althea root,[1] and has observed this medicine to diminish considerably the trembling, uneasiness, formications, and feeling of weakness. He has found opium useful, especially when given with the view of checking the formications, twitching of the muscles, cramps, and convulsions; but he has not noticed this drug to relieve the patient from the nightly hallucinations, and considers it as contra-indicated

---

[1] The prescription used by Dr. Huss is as follows:
℞ Fermentolei solani, ℈j;
Radix althææ pulveratæ, ʒj;
Misc. 'c. Syrupi althææ, q. s., ut f. mass. e-qu. form. pilulæ æq. N. xl.
Two pills are to be taken six times daily.

in cases attended with delirium during the daytime. He alludes to camphor as having the property of allaying the nervous uneasiness, tendency to delirium, and the occurrence of hallucinations while endeavouring to fall asleep; he also recommends camphor as a means of checking the giddiness and faintness; it is to be given in doses of from one to five grains. He has obtained good results from the use of Arnica (Flores arnicæ) when, after having been cured of the trembling and formication, the patient still complained of a feeling of weakness, accompanied with dulness of the mental faculties, noises in the ears, and the appearance of flying specks when in an erect position. He has recourse to medicines containing iron when others fail in their action. Finally, Huss often makes use of spirituous fluids, prescribing a glass of brandy to be taken twice daily, or a daily allowance of two glasses of port or as much sherry, or three or four table-spoonfuls (esslöffel) of Tinct. Absynthii or Tinct. Cinchonæ; he prescribes sometimes forty or sixty drops of Spiritus Etheris Sulphurici (ether spirituosus) to be taken twice or three times a day, or half a bottle of porter to be drunk in the forenoon. I have tried the effects of opium, carbonate of ammonia, preparations of iron, bitters, and other medicines, which were attended with more or less benefit; and in cases where the digestion was disordered, opium has been administered with very good results, besides my finding it very useful in bringing on sleep in one or two cases where oxide of zinc had failed to produce this result. I have not often observed bitters and iron to be beneficial at the outset of the disease, but the patient being in a fair way of recovery, steel and quinine have proved of great service. In some cases attended with headache, considerable relief was obtained from the application of a small blister behind the neck.

If, however, chronic alcoholism be considered as depending on a peculiar diseased condition of a certain part of the body, owing to the action of a poison, no remedy can be looked upon as decidedly efficacious unless it exerts its power not directly on the symptoms themselves, which are but the signs of the illness, but on the principle of the disorder. Bearing this in mind, I have endeavoured to discover a treatment which, by acting immediately on the nervous system should remove its diseased condition—the result of the long-continued abuse of alcoholic stimulants, thereby acting as a means of arresting the symptoms of the illness. I am consequently not about to recommend one remedy for a certain symptom, and another remedy for another symptom, but shall endeavour to show that there exists a substance, possessed of powerful and definite medicinal properties, and having the remarkable property of restoring to health, or at all events of greatly relieving the disordered nervous system of persons suffering from chronic alcoholism; the medicinal agent in question acting efficaciously in cases where the principal symptom may be either sleeplessness, or hallucinations, or trembling, or any other; and this substance is OXIDE OF ZINC.

I shall first proceed to give an account of the action of oxide of zinc on the human body in health and in disease, and then show

how powerful an agent it is for the cure of the particular complaint of which we are treating.

PHYSIOLOGICAL PROPERTIES OF OXIDE OF ZINC.

This subject has been ably investigated by Dr. Michaelis, of Tubingen, Dr. Bouchut, De la Roche, Dr. Herpin, of Geneva, and others.

Oxide of zinc, although very sparingly soluble in water, is readily dissolved in the stomach and intestines. According to the experiments of Michaelis, the metal finds its way into the blood, bile, and urine; twenty-four hours after the injection of a salt of zinc into the crural vein the bile exhibited evident traces of zinc. It appeared in the bile previous to being eliminated from the body with the urine. This gentleman also observed, from experiments upon animals, that large doses of oxide of zinc produced erosions and ulcerations of the mucous membrane of the stomach, and that it may induce within the organs of respiration granulations analogous to miliary tubercles. Moreover, according to Michaelis, the long-continued use of moderate doses of oxide of zinc interferes with digestion, producing anæmia and marasmus. He found the blood of dogs taking oxide of zinc to contain 0·99 or very nearly 1 per 1000 of fibrin, instead of 1·92, the normal proportion. Dr. Bouchut, on repeating the experiments of Michaelis, succeeded, however, in giving to a strong rabbit doses of from five to ten grains of oxide of zinc without inconvenience to the animal.[1]

Dr. Herpin, after a long and careful series of observations on the effects of oxide of zinc upon the human body, has arrived at the following conclusions:[2]

That it is a perfectly harmless remedy, and may be given in doses of as much as six grammes (90 grains) a day, for a very con-

[1] 'Etudes sur le Lactate de Zinc dans l'Epilepsie,' par le Dr. Herpin.
[2] 'Du prognostic et du traitement curatif de l'Epilepsie,' 1852, p. 565.

siderable time, without producing any other inconvenience than temporary uneasiness.

That its physiological effects are confined to a mild action (action légére) on the intestines, consisting usually, in the case of adults, of nausea which may occasion vomiting; and, in the case of children, of slight diarrhœa.

That the medicine is easily made to be tolerated without discomfort, by beginning, in the case of adults, with 4·5 to 6 grains a day, and in the case of children with from 1 to 2·5 grains a day, these quantities being divided into three or four doses; and then giving, every week, from 2·5 to 3 grains more daily to adults, and from 1 to 2·5 grains more daily to children.

That the form of pill is sometimes a means of enabling its being tolerated.

Finally, that the uneasiness occurs less frequently when the medicine is taken an hour after a meal, than when fasting, and that the first dose in the morning is always that which is attended with the most discomfort.

It appears that since Dr. Herpin wrote his

valuable treatise on epilepsy, he has slightly modified his opinion respecting the first of his conclusions, for, in a paper entitled 'Etudes sur le Lactate de Zinc dans l'Epilepsie,' published by him in 1855, he observes, that after a long-continued use of oxide of zinc, when it has been taken to the extent of from 120 to 473 grammes, unfavorable symptoms may occur, young women being more especially subject to suffer in such cases from anæmia and chlorosis; the symptoms, however, only acquiring some degree of importance, when the treatment has been persevered in for one month after the first appearance of the unfavorable effects.

I have confined myself in my researches on the action of zinc on the human body, to the use of the metal under its form of oxide, and have obtained the following results:

1. That after taking this substance in doses of from two grains and upwards in the case of adults, a feeling of nausea is sometimes perceived, but seldom to the extent of producing vomiting. This effect is diminished if,

according to Dr. Herpin's practice, the medicine be given about an hour after a meal.

2. That after persevering with the treatment for some days, the medicine is in most cases tolerated, and the nausea and uneasiness produced at first, diminish, and even disappear.

3. That a slight giddiness attended with the appearance of black specks before the eyes, and rumbling noises in the ears, may accompany the nausea occasioned by oxide of zinc; this is an indication of the doses being too high, and on diminishing them, these symptoms disappear.

4. A very important and remarkable effect of oxide of zinc is the power it frequently possesses of producing sleep.

5. I have not noticed the long-continued use of oxide of zinc to produce evidently deleterious effects, even after it has been taken for a considerable length of time.

The feeling of nausea and sickness occasioned by preparations of zinc appears to vary, to a certain extent, according to the form of the compound, for we are informed by Dr.

Herpin that *lactate* of zinc is not so likely to be liable to this inconvenience as the *oxide*. The fact of oxide of zinc producing less uneasiness when taken after food is obviously owing to its state of dilution in a full stomach. Dr. Herpin has also observed that oxide of zinc is better tolerated when taken under the form of a pill, which may be accounted for by assuming that the substance is dissolved in the stomach, under that shape, less rapidly than under that of powder. It must be remembered, however, that oxide of zinc made up into pills may altogether escape absorption, and for that reason I prescribe it frequently as a powder. Some years ago, a female out-patient of the Westminster Hospital, who was taking pills of oxide of zinc and confection of roses, brought me a hard concretion she had removed from her motions. I found it, on examination, to be one of the pills taken, which had consequently escaped absorption. It is very questionable whether the whole of the medicine is absorbed when given in large quantities, for it is remarkable to what extent the doses may be increased with impunity; I have frequently prescribed as much as twenty grains of the substance to be taken twice a day, and in two cases of epilepsy the dose was raised to thirty-five grains twice a day. With respect to the mechanism of its absorption, it is first dissolved, not only by the gastric juice, but also by the free acid (probably lactic acid) and acid phosphates of the juice of the meat contained in the stomach; the fats of the meat taken also very probably combine with it, especially the fatty acids derived from the neutral fats of food; the conversion of more or less fat into fatty acids being a phenomenon I have shown to take place invariably during digestion,[1] and both M. Chevreuil[2] and MM. Jeannel and Moncel,[3] have observed that fats enter into combination with metallic oxides, forming a peculiar kind of soap.

It is not a little remarkable that oxide of zinc should in some cases produce the very symptoms it is intended to cure, namely,

[1] 'Proceedings of the Royal Society.' June, 1858.

[2] Chevreuil 'Sur les Graisses.'

[3] A paper read to the Academie de Médecine, Nov. 3d, 1857.

giddiness and faintness; and for this reason it should not be given indiscriminately. As a general rule I have found it objectionable in chlorosis, and with females of a weak constitution. In some cases of hysteria I have also been obliged to withhold it, from its being decidedly objectionable; the following is a case in point. C. G——, aged 26, complains of headache and giddiness. Previous to her marriage, six years ago, she had been subject to fainting fits, but since then the affection had not returned. She was treated with small doses of oxide of zinc, and about five minutes after taking the first dose she felt very sick, and fainted. After taking a second dose, she fainted a second time, remaining unconscious for two or three minutes on each occasion.

Sometimes, though rarely, an apparent increase of the existing symptoms will occur in cases of chronic alcoholism treated with oxide of zinc; but on diminishing the dose, the unfavorable symptoms at once disappear.

The property of oxide of zinc of frequently producing sleep—a power it appears to possess exclusive to all the other metals—is very remarkable. I have observed this phenomenon, not only where oxide of zinc was given for the treatment of chronic alcoholic intoxication, but also when administered in other cases. A gentleman, taking oxide of zinc for chorea, complained of his feeling so drowsy after dinner that he was obliged to go to sleep every day at that time, greatly to his discomfort, as he boarded with a family, and was much annoyed at this apparent breach of sociality. W. S——, aged 11, treated with oxide of zinc for chorea, experienced great drowsiness every evening at half-past eight o'clock, although he had never before felt sleepy at that hour. Another male patient, aged 52, taking oxide of zinc for the treatment of vertigo and headache, stated that since he had begun the powders he felt very sleepy, and could sleep all day long. W. J——, aged 75 (Case 5), taking oxide of zinc, is reported as sleeping from one till three o'clock, p.m., although he had never slumbered so long before in the daytime. E. B—— (Case 6), became very sleepy in the daytime since taking the powders. W. H—— (Case 7), treated with oxide of zinc, has observed that he feels very sleepy in the daytime; when sitting to read he falls asleep; he was never in the habit of sleeping in the day before taking the powders. J. W— (Case 11), states that the powders of oxide of zinc make him feel drowsy; he falls asleep about one hour after taking them, and sleeps for an hour; had not previously been accustomed to sleep during the daytime.

From the frequent necessity, in cases of chronic alcoholism, to continue giving oxide of zinc for a comparatively great length of time, it is important to make sure that this medicine cannot act as a slow poison, such as many other metals do, for instance mercury and lead. Indeed, as previously observed, it would appear, from Dr. Herpin's experience, that unfavorable symptoms do show themselves in certain cases, from a protracted treatment with this metallic oxide. I have observed, however, that when the medicine acts unfavorably, as is sometimes the case with young and weak females, and especially those suffering from chlorosis, the patients are at once very quickly affected by the drug, offering, consequently, no opportunity of examining the result of a long-continued treatment. I have given oxide of zinc in a great number of cases for several months, without producing any dangerous, or even *evidently* inconvenient symptoms. The following are instances of this kind.

M. R——, aged 24, admitted under my care as an out-patient of the Westminster Hospital, in October, 1855, is suffering from epilepsy. She began the treatment on the 27th of October, with six grains of oxide of zinc twice a day, the dose being rapidly increased till the 15th of December, when she took no less than thirty-five grains of oxide of zinc twice a day. My notes of this date state:—Had a fit on the 13th, which was not severe; looks pale and feels giddy. The symptoms were probably owing to the fit the patient had two days previously, and not to the zinc; the dose of thirty-five grains was perhaps rather large, but as will subsequently be seen, if it produced paleness and giddiness, these symptoms were but transitory. On January the 16th the dose had been gradually reduced to ten grains. By the 9th of February it had been again increased to twenty-two grains twice a day. On the 20th of February, she was taking fifteen grains twice a day, and she continued with that quantity till the 5th of April, having taken large doses of oxide of zinc for four months and a half. She had evidently benefited from the treatment, as on the 2d of April the following note of her case occurs in my note-book:—Continues to feel quite well, no return of the fits. In the next case the patient continued the treatment with oxide of zinc for a much longer period. H. J——, aged 35, admitted as an out-patient on the 26th of May, 1855, and suffering from epilepsy. He began, on the day of his admission, with one grain of oxide of zinc twice a day, which was increased gradually till the 5th of December, when he took thirty-five grains twice a day; on the 12th of December, I recorded the following state of his case:—Feels quite well, no return of fits; showing at least that this enormous dose of oxide of zinc had produced no ill effect. He continued taking from fifteen to twenty grains of the drug, with an interruption of a few days (when sulphate of zinc was tried), till the 5th of July, having been for nearly fourteen months under treatment with oxide of zinc, and, at all events, not the worse for the medicine.[1]

The following case is less remarkable, as the oxide of zinc was not taken for so long a period as in that of H. J——. It is, however, well calculated to show that this medicinal agent does not act as a slow poison.

R. M —— admitted as an out-patient of the Westminster Hospital, on January 16th, 1856, suffering from incessant trembling of the right arm. This patient began with one grain of oxide of zinc twice a day, the dose being gradually increased till the 24th of February, when he was taking fourteen grains of the oxide twice a day. On the 1st of March ten grains were prescribed for a dose, which he continued taking till the 14th; after this date the dose was gradually raised to twenty grains; this he continued taking from

---

[1] I remember having heard this patient complaining of slight loss of memory, but it was not possible to ascertain whether this symptom depended on the epileptic fits or on the oxide of zinc.

the 19th of April till the 16th of May, having thus been treated with large doses of oxide of zinc for four months. The trembling had much diminished under this treatment, and no new symptom had occurred; consequently the medicine had certainly not produced any injurious effects.

Many more cases might be reported, if necessary, showing that the long-continued use of oxide of zinc as an internal remedy, is attended with no evident evil results.

THERAPEUTICAL PROPERTIES OF OXIDE OF ZINC.

Oxide of zinc may be considered as a tonic for the nervous system, and a sedative and antispasmodic. The other compounds of zinc used in medicine appear to be possessed more or less of the same therapeutical properties as the oxide. Dr. Barnes has administered phosphate of zinc with favorable results in cases of epilepsy. The sulphate has been employed with advantage by Dr. Hughes, Dr. Addison, and Dr. Barlow for the treatment of chorea, and the lactate has been used in epilepsy by Dr. Herpin.[1]

Dr. Golding Bird has mentioned in the wards of Guy's Hospital that zinc has a peculiar and specific influence on the nervous system, in about the same manner as iron on blood.[2] I have heard it asserted, however, that this substance is possessed of medicinal power merely on account of the faith the patient places in its efficacy, meaning, in plain language, that it exerts no action whatever. As a proof of the effects of oxide of zinc not arising from its influence on the imagination, I may perhaps be allowed to report, in addition to Dr. Golding Bird's valuable opinion, the case of a little dog treated successfully with this drug. On the 2d of November, 1855, a small spaniel, kept in my

---

[1] Dr. Theophilus Thompson has found oxide of zinc very useful for the removal of colliquative sweating in phthisis.—'Lancet,' 1854, vol. i.

[2] 'Lancet,' 1851, vol. i.

house, was suddenly seized with a peculiar and uninterrupted moaning. The animal was very restless during the night, and the next morning had a fit attended with violent screaming and barking, which lasted about one minute; it was followed by other similar attacks, which increased in frequency until there was hardly any intermission between them, the animal running round and round from left to right, and crying out most piteously. The fits continued the whole of the succeeding night. On the 24th, the violence of the attacks had increased; there was partial paralysis of the posterior extremities; but towards the evening the paroxysms diminished in violence and frequency. On the 25th they were hardly perceptible, and on the following day the animal had resumed his lively character. Since then, with the exception of two or three epileptiform fits, this little dog has had no return of the illness. I began the treatment in the present instance with opium; the dog was made to take ten drops of laudanum three times during the first day, and then twenty drops at night, but without stopping the fits. I next made him inhale chloroform until he became perfectly insensible, which had no further effect than that of diminishing the fits at the time. On the 24th I gave him four grains of oxide of zinc in two doses. (One grain had been given to him on the 22d); the same on the 25th and the 26th, when the treatment was concluded. This case proves beyond doubt that oxide of zinc does not cure by acting on the imagination, but that it is really in itself a powerful remedy.[1]

Oxide of zinc exerts a remarkable action in certain cases of functional derangement of the nervous system. I have administered it in epilepsy, chorea, cases of mild hysteria, paralysis and lead palsy, exhaustion from excessive mental work, and chronic alcoholic intoxication. I succeeded in obtaining favorable results in certain cases

---

[1] I have not found it stated in my notes at what time on the 24th the oxide of zinc was taken; but recollect distinctly giving the remedy as a last resource, thinking the dog was then dying; the treatment had consequently been commenced at the most acute stage of the disease.

of epilepsy and chorea; indeed, I had been induced, at first, to consider this agent as a specific for the treatment of epilepsy, and under this impression had communicated to the 'British Medical Journal' (for the 23d of November, 1855), a report on the use of oxide of zinc in epilepsy. Since that time, however, having continued to prescribe the remedy for the treatment of this disorder, I am now obliged to admit that it seldom, if ever, cures the disease, although it is certainly often attended with beneficial effects. I have obtained similar results in chorea; indeed oxide of zinc is perhaps more effectual in chorea than in epilepsy. Some cases of mild hysteria were rapidly cured under a treatment of oxide of zinc, in others it produced scarcely any effect, and in certain instances the medicine proved objectionable. In the following cases oxide of zinc acted most favorably. M. J—, aged 48, admitted on November 16th, 1855. She suffered two years ago from an abscess in the forehead, and has complained ever since that time of constant headache. Is very subject to giddiness and fainting, but never had any fit: muscæ volitantes and tinnitus aurium, no bolus hystericus. Sleeps very restlessly, appetite pretty good; has not menstruated for a year. On admission, one grain of oxide of zinc was ordered to be taken twice a day, which on the 30th had been gradually increased to six grains twice a day. On the 19th and 21st she had half an ounce of castor oil. On December the 7th I took the following note: Is now quite well and strong, no more headache or giddiness, sleeps better, though not yet quite soundly. Is dismissed. A. M——, admitted on the 28th November, 1855. Complains of headache; muscæ volitantes, and tinnitus aurium for the last six weeks; appetite good, bowels rather confined; she was treated with oxide of zinc, and dismissed cured on the 19th of December. I have not found paralysis and lead-palsy to be relieved by the internal use of oxide of zinc.

Mental exhaustion.—Remarkably favorable results have been obtained from the prescription of oxide of zinc in cases of cerebral affection arising from excessive mental exertions;

and this disease, which is in many cases difficult to cure, is rapidly brought under control by the above means. I have treated four cases of this description by means of oxide of zinc. The first is that of Mrs. B—— (see p. 57). I was requested to see her on the 21st of November, 1855. She had been for the last three months almost exclusively engaged in literary pursuits, and ascribed her illness to excessive reading and exertion of the mind. The patient was treated with one grain of oxide of zinc twice a day, and a blister was directed to be applied behind the neck; castor oil was given once. On the 23d of November I found her much better; on the 24th there was a slight relapse, which soon gave way under the same treatment. She expressed herself quite recovered on the 26th. I again visited her about a fortnight afterwards, and found her quite well.

The next case is of considerable interest, as, after other means had failed, the patient recovered under a treatment with oxide of zinc. Miss E——, aged 15, has been for the last three years a pupil at a public school, and has applied herself with excessive zeal to her studies. She suffers from headache and giddiness, occasional tinnitus aurium and muscæ volitantes; mental abstraction become difficult. I advised her to give up her studies, and began the treatment with quinine and iron, but without benefit; one grain of oxide of zinc was then ordered to be taken twice a day, and a small blister directed to be applied behind the neck; the same part was again blistered a fortnight afterwards. The dose of the medicine was increased gradually until she took three grains twice a day. After taking sixty pills of oxide of zinc she had completely recovered, nothing more than an occasional and but slight headache being left, felt especially when confined in a room or a crowded carriage.

In the following case excessive mental exertion was also the exciting cause of the complaint, the individual being in other respects predisposed to it. Again oxide of zinc proved most useful. J. R——, aged 19, admitted an out-patient of the Westminster Hospital on the 23d of November, 1855. About two years previous to admission had a fit, subsequent to a fall on his head from a height of four feet. The accident occurred in the afternoon and the fit took place on the evening of the same day, lasting till the middle of the following day; he was quite insensible for nearly the whole of that time; has had no return of the fits since then. Is a numerical printer, and besides his business has been very busy studying Greek and Latin at night, which he believes to have been the main cause of his present illness; he dropped his literary pursuits a few days ago, but continues attending to his business. For the last two years has been suffering from headache, pain in the region of the heart, and palpitation. He trembles at any sudden and unexpected disturbance, and occasionally feels a sensation of fainting. Was ordered to take two grains of oxide of zinc twice a day. On the 30th he was much better; the dose was increased to five grains. He returned to me for the last time on December the 7th, being greatly improved in every respect.

It may be well here to remind the reader of the melancholy case of W. T— (page 58). This patient recovered completely under a treatment of oxide of zinc.

Dyspepsia.—The following case is interesting, as showing that, in some cases of inveterate chronic gastralgia, oxide of zinc may effect, or assist in effecting, a perfect cure. I called on Mr. R— on the 1st of October, 1856. He was suffering from great pain in the epigastric region, occurring three times a day, about two hours after each meal. The pain was attended with a sensation of great weakness and mental depression; he felt quite well in every other respect. I treated the case with quinine and sulphuric acid; rhubarb and sulphate of magnesia; iron and quinine; and compound soap pill. On the 22d of November he was better, though still complaining of the pain. In January, 1857, this gentleman again requested me to see him; he was suffering from the same symptoms as previously. On examining his urine, on the 3d of February, I found it to contain uric acid and urates, and prescribed ten minims of liquor potassæ, to be added to his beer at dinner, and ordered one grain of oxide of zinc, made into a pill with confection of roses, to be taken twice a day. On the 7th of February there was a decided improvement. On the 12th the improvement had continued; he was directed to take the liquor potassæ twice a day, and the dose of oxide of zinc was increased to two grains. On March the 7th he only felt a slight pain in the stomach every two or three days; he had been taking four grains of oxide of zinc for a dose. Was directed to omit the liquor potassæ, and take five grains of oxide of zinc twice a day. The following note was recorded on the 27th of March: Has had no return of pains for the last eight or ten days; has taken no oxide of zinc for a week. I then discontinued attending, but have heard within the last few days that since March, 1857, or for more than two years, there has been no return of the dyspeptic symptoms.

I now propose returning to our subject, and describing the action of oxide of zinc in cases of chronic alcoholic intoxication.

CHRONIC ALCOHOLISM TREATED WITH OXIDE OF ZINC.

Let it first be well understood, that it is not advisable to give oxide of zinc indiscriminately in every case of disturbance of the nervous functions owing to intemperance. The beneficial results of this treatment are more especially observed in those cases of chronic alcoholism which are unattended by any organic disorder, and when the disease occurs in an individual of a sound constitution. The various affections of the lungs are peculiarly liable to diminish, in a considerable degree, the property of oxide of zinc of controlling the illness in question. Rheumatism may next be mentioned, but I must not omit to observe that chronic alcoholism is very frequently attended with a symptom resulting directly from the alcoholic poisoning, and which might be mistaken for rheumatism. It is described by the patient as a weakness and pain across the hips, and weakness in the knees, preventing a heavy weight being carried, or even interfering with the act of walking. This symptom usually gives way under a treatment with oxide of zinc, but may sometimes remain after the patient has recovered in every other respect; in some cases it will eventually disappear by having recourse to tonics. Gastric disorders, which so frequently accompany chronic alcoholism, are sometimes rapidly subdued by oxide of zinc; this occurs, no doubt, principally in those cases where the gastric symptoms depend entirely on the disturbance of the functions of the nervous system; and that such a disorder does actually exist is shown by the following observation of Dr. Budd, in his book on 'Diseases of the Stomach.' Having mentioned the organic changes occurring in the stomach in cases of indigestion of drunkards, he states (page 287)—"But the gastric disorder we are considering may occur without either of these events, and when the stomach in case of death may exhibit no striking marks of disease." When the gastric symptoms are very urgent, such as great pain in the stomach with much vomiting, I am in the habit of prescribing opium at night, in the form of compound soap pill, independently of the oxide of zinc. In some of these cases the substitution, for a few days, of carbonate of ammonia for oxide of zinc was attended with benefit.

The following is the usual effect of oxide of zinc in simple cases of chronic alcoholism: First, the sleep is improved, the patient does not lie so long awake at night, and the nightmare becomes less frightful; then, the hallucinations decrease, the patient is no longer troubled with black specks passing constantly before his eyes, or with the sight of imaginary objects, such as insects or other animals crawling about the room, and extraordinary noises are no longer heard; the attacks of trembling also diminish in frequency if not in intensity, and gradually pass off. This improvement is attended with an increase of appetite, as well as a marked diminution of the gastric symptoms; and when the patient can take food and digest it well, he may be looked upon as in a fair way towards recovery. Gradually, muscular power returns, and the mental depression, which frequently accompanies chronic alcoholism, disappears; the patient becomes cheerful and happy, and expresses with gratitude his joy at feeling quite well.[1] When the disorder is complicated with an organic disease, I have found it advisable to begin with oxide of zinc, in order to alleviate as much as possible the functional derangement of the nervous system, and then to adopt such a course of treatment as may be considered the most suitable to the occasion.

The following cases have been selected as well adapted to illustrate the action of oxide of zinc in chronic alcoholism, and will be found worth attention. They are divided into two classes, the first class including those cases in which the disease assumed its simple form, and the second including those which were accompanied with other disorders. Each class of cases is also divided into two subdivisions: the first comprehending patients treated for chronic alcoholism at the time they were over-indulging themselves in spirituous beverages; and the second those who came under treatment some time after having given up the habit of drinking.

[1] It is to be understood that during the period of recovery the symptoms are frequently not relieved in the above-mentioned order.

CLASS I.—DIVISION I.

Cases of Chronic Alcoholism not complicated by other diseases, and occurring during the period of intemperance.

W. B—(Case 1), aged 34, a carpenter, admitted as out-patient at the Westminster Hospital, on November the 21st, 1855.

Has been addicted from his youth to the excessive use of alcoholic liquors, and in the habit of taking one pint of gin daily, with two or three pints, and occasionally six or eight pints, of beer. Has had a slight attack of delirium tremens. Cannot sleep at night, but feels composed if he has recourse to spirituous libations before going to bed. Ten days ago he drank more than usual, and now suffers from a peculiar and very unpleasant sensation of choking, from giddiness, great weakness in the legs, and trembling of the body. Since the 10th of November, has reduced his allowance of alcoholic beverages to four pints of ale and one quarter of gin. On the 21st, was ordered to take two grains of oxide of zinc, in the form of a pill with confection of roses.

On the 24th, there was already a very great improvement; had slept well the night before. No more trembling of the body, although the tongue remains slightly tremulous; appetite better, and is himself surprised at his improved condition. Zinc. Ox., gr. v, bis die.

27th.—Has not slept so well since the 24th. Still complains of uneasiness in the throat, and is very subject to flatulence. Zinc. Ox., gr. viij, bis die.

December 1st.—The recurrence of unfavorable symptoms has discontinued, and he feels much better; the sensation of choking is not nearly so strong. Zinc. Ox., gr. x, bis die.

8th.—Improvement continued, sleeps well, appetite good; takes one pint of beer and half a quartern of gin, daily. Zinc. Ox., gr. xij, bis die. (From the 26th of December, the oxide of zinc was given under the form of powders instead of pills.)

He continued taking the same medicament until the 12th of January, 1856. On the 18th December, fourteen grains had been prescribed to be taken twice a day, and a small blister ordered to be applied over the larynx, with the view of relieving the sensation of spasmodic dyspnœa; this was attended with some slight benefit.

On the 5th of January, the dose of oxide of zinc was increased to fifteen grains twice a day, and another blister was applied to the larynx. On the 12th, he is reported as still suffering from uneasiness at the glottis, although otherwise quite well. He was then ordered to give up the oxide of zinc, and take a mixture containing rhubarb, &c. On the 19th, having observed that he used to swallow a large quantity of air, I advised him to give up that habit, and on the 26th the spasmodic dyspnœa had greatly diminished. On the 30th, is nearly free from this last symptom, sleeps quite well at night, and enjoys a good appetite; takes three pints of beer a day and no spirits. Dismissed cured.

This unfortunate man returned to me on the 21st of January, 1857, having again taken to drinking, and that shortly after he had left off attending the Westminster Hospital. He had also again contracted the habit of swallowing air, and suffered much from spasmodic dyspnœa. He had resumed his old habit, drinking from three to four pints of porter, with an occasional glass of gin, and gradually this had been increased to from four to five quarts of porter daily, and a quartern of gin. Has had no attack of delirium tremens, but suffers at times from great giddiness. He awakes very often at night with difficulty of breathing; is very low in spirits; appetite good; no hallucinations. Has been under medical treatment at Lewisham, but without obtaining any relief. To take two grains of oxide of zinc, twice a day; and he was reminded of the importance of his giving up the habit of drinking.

24th.—Sleeps better at night, is in better spirits, has diminished his allowance to two pints of porter daily. Zinci Ox., gr. iij, bis die.

28th.—Improvement continues, drinks three pots of beer daily and a quartern of gin, endeavours to swallow less air, dyspnœa much diminished, does not suffer from flatulence so much as before. Zinc. Oxy., gr. v, bis die.

February 4th.—He now sleeps very well at

night and is on the whole much better. Ox. zinc., gr. viij, bis die.

14th.—On the 4th he indulged himself a little in his old habit of drinking, and the following night suffered from sleeplessness and a return of dyspnœa. Zinc. Ox., gr. xij ; Pil. Saponis co., gr. iij, hora somni quotidie sum.

25th.—Is again better; to continue the powders.

March 6th.—From the 28th has returned to the habit of drinking; great giddiness, no sleep at night, loss of appetite, and much mental depression.

My notes of this case end here; W. B— may have attended the hospital some little time longer, without deriving any permanent benefit from the treatment; as he never gave up, and, in all probability, never will give up the habit of drinking.

T. S— (Case 3), aged 33, an engineer on board a steam-boat, admitted as out-patient on the 2d of April, 1856.

Is in the habit of taking about one pint of spirits and four or five pints of beer daily, and has suffered from several attacks of delirium tremens, the last occurring about four months ago. He complains at present of want of sleep, giddiness and much nervous uneasiness. Tongue slightly tremulous; much pain in the epigastric region; feels in the throat a peculiar sensation as of choking, especially in the morning, which he is in the habit of relieving by having recourse to drink, also complains of great weakness. Zinci Ox., gr. ij, bis die.

April 4th.—Slept very well last night, feels less uncomfortable; the sensation of choking has completely disappeared. Has reduced his allowance, first to two pints of ale and two glasses of gin, and then to two pints of ale without gin. Zinc. Ox., gr. iv, bis die.

12th.—No return of uneasiness, sleeps now quite well, does not suffer from spasmodic dyspnœa, but feels rather weak; he says he has taken a great deal of medicine, but felt no relief until I prescribed for him powders of oxide of zinc. Has returned to work and drinks about two pots of beer and two or three glasses of gin daily. Dismissed cured.

G. P— (Case 4), aged 35, admitted as an out-patient on the 17th of May, 1858.

Has been addicted for a considerable time to the excessive use of alcoholic liquors. Six years ago he went to America, and spent there two years in the army, during which time he drank a great deal of brandy and water. Has been in the constant habit, for many years, of drinking about eight pints of beer daily, and also occasionally pure brandy. Had two attacks of delirium tremens within the last two years, on each occasion the attack lasted three or four days. On the 12th of May, according to his own expression, had a little too much drink, and felt unwell the next morning; since then he suffers from paroxysms of great trembling, cannot sleep at night (which is, however, a symptom of long standing), and is subject to dizziness of sight although not to headache. On the 15th, as he was walking in the street, he fancied there were ropes dangling about him; complains of slight spasmodic dyspnœa. Appetite bad, occasional pain in the region of the bladder. Zinc. Ox., gr. ij, bis die. To take two pints of beer daily, and no more.

24th.—Feels and looks better, trembling now very slight, sleeps very well at night. Began sleeping well the first night after taking the powders. No hallucinations; no more dyspnœa; still feels weak. Has taken the powders regularly, and adheres to two pints of ale daily. Zinc. Ox., gr. iv, bis die.

27th.—Improvement continued, no more trembling, sleeps well at night, appetite pretty good. No more pain in the region of the bladder, slight constipation; continues taking two pints of beer daily. Left off attending.

E. B—— (Case 6), aged 40, a sailor, admitted as out-patient on the 10th of June, 1858.

Ten or eleven years ago this patient contracted the habit of drinking brandy to the extent of about a dozen glasses a day, which he continued doing for eighteen months. During that period he was occasionally intoxicated, but used to get over it so well as to be able to resume his work, and to drink again the following morning. He admits having suffered from an attack of delirium tremens in 1843, although not then in the habit of drinking to excess, and had indulged himself, on that occasion, to a considerable extent in alcoholic liquors for four or five days. Eighteen months ago he reduced his allowance of rum from four to two glasses a day, taking beer occasionally to the extent of five or six pints when he could go ashore. Is subject to pain in the loins, and sometimes in the stomach; tongue usually pretty clean; suffers now and then from headache, giddiness, and tinnitus aurium. Often sees a shadow passing before his eyes and then rapidly disappearing; is restless at night, and cannot sleep well. When about twenty years old had four series of fits, probably epileptic, but from the age of twenty-two had no return of those attacks; sometimes his legs tremble, they are very weak, especially the right one. In February last, applied for medical advice and obtained relief, although he never recovered the free use of his right leg, and the left remained weak. Zinc. Ox., gr. ij, bis die.

14th.—Is stronger, but complains of the

powders occasioning sickness, feels very sleepy in the daytime, sleeps well at night.

17th.—Improvement continues. Zinc. Ox., gr. ij, bis die.

24th.—To continue the powders.

28th.—Sleeps very well at night, no longer any trembling, or headache, right leg not yet regained its whole strength; drinks now daily two pints of porter and takes a glass of gin occasionally. P. pulv.

July 8th.—The powders make him very sick, and he has vomited after breakfast. Zinc. Ox., gr. iij, bis die.

19th.—Complains of a weakness in the right knee, the leg occasionally giving way. In other respects is perfectly well. Advised him to take some sea-bathing, or at all events bathe in cold water, and give up every other medical treatment.

J. P— (Case 8), aged 39, a coal-porter; admitted as out-patient on the 11th of October.

For the last twenty years has been accustomed to drink five pints of beer and a glass of gin daily. He is, moreover, intoxicated once a week—on Saturday night. Has been in bad health for the last fifteen months. On sitting, his legs tremble, but they remain quiet when he stands. Sleep very restless, is incessantly turning over at night, complains of giddiness and shooting pains across the temples, and occasionally a fog or cloud passes before his eyes. No tinnitus aurium. Is frequently sick when getting up in the morning; very weak in the knees, and suffers from an acute pain in the right hip. I directed him to take one pint of ale daily. From the date of admission till the 28th of October this patient was not treated with oxide of zinc, and obtained no relief. On the 28th of October I prescribed for him two grains of oxide of zinc to be taken twice a day.

November 4th.—Great relief since he began the powders, sleeps pretty well at night. No longer any nightmare, not so much trembling in the legs or sickness in the morning, feels a little stronger, but still complains of weakness, especially in the right hip. Passing shadows no longer perceived. Zinc. Ox., gr. iv, bis die.

11th.—Complains of a slight cough, knees and hips still weak, in other respects quite well. His legs to be rubbed with turpentine liniment, and to take a mixture of iron and quinine.

On the 22d he returned to work, but on the 29th, the weakness continuing, I directed him to resume the oxide of zinc—gr. iv, twice a day, and omit the mixture.

On the 6th of December feels much stronger, and now goes to work regularly.

On the 16th, as he did not sleep quite so well, three grains of compound soap pill were prescribed to be taken every night.

On the 23d he gave up attending the hospital, having quite recovered.

J. W— (Case 11), aged 18, admitted as an out-patient on the 28th of October, 1858; a hawker.

Has been for five years of intemperate habits, although he has considerably reduced his allowance of drink for the last three weeks. When in company of friends, that is, from once to three times a week, takes occasionally as much as eight pints of beer and one or two glasses of gin. On the 7th instant he drank about fifteen pints of beer and five or six glasses of spirits. The next day he was seized with great nausea, giddiness, and trembling all over the body. Besides these symptoms he also suffers from a pain in the left hip and great weakness in the legs, and does not sleep well at night. To take one pint of beer daily and no spirits, and two grains of oxide of zinc, twice a day.

November 8th.—The powders have made him sleep in the daytime; he does not, however, sleep soundly at night; the pain has shifted to the right hip. No longer any giddiness, but trembles very much in the morning for about five minutes only. He is very weak, and thinks he could not carry a quarter of a hundred weight. Zinc. Ox., gr. iv, bis die.

11th.—P. Pulv. the hips to be rubbed with compound soap liniment.

18th.—No longer any trembling whatever; he would feel quite well, were it not for the weakness of the hips. To take a mixture of iron and quinine.

22d.—No improvement from the last medicine; to resume the oxide—gr. iv, bis die.

23d.—Returned to work (pushing a heavy wheelbarrow) yesterday evening, for about two hours; felt no inconvenience from it except stiffness.

25th.—Is stronger, pushing his wheelbarrow for three hours. Zinc. Ox., gr. iv, bis die.

December 6th.—Has not attended since the 25th ult., can now work as he used to do before he fell ill; pain in the hips very slight, and felt only occasionally. Discharged cured.

CLASS I.—DIVISION II.

Cases of Chronic Alcoholism not complicated by other diseases, and occurring in patients having resumed habits of sobriety.

G. R— (Case 18), aged 38, a general dealer; admitted December 20th, 1858.

Has been addicted to the habit of drinking from youth, till eleven months ago, taking from three glasses to a pint of spirits daily, and has had as much as nearly two pints of spirits in one day; his daily allowance of ale has been four or five pints; never suffered from delirium tremens or any other disease, but gave up drinking completely eleven months ago, because he found it did not agree with his health. The symptoms he is now suffering from have lasted since then; these are, great giddiness and headache, passing clouds, and sudden blindness for two or three minutes, but no actual hallucinations. Cannot sleep at all at night. Trembles a little all day long, especially when moving about. Appetite very bad. No pain in the stomach. Has an eczematous eruption on the face. Has done no work since the 17th.

On the 23d December he was ordered to take two grains of oxide of zinc twice a day. He had so far improved on the 27th as to feel capable of returning to work; he suffered no longer from giddiness and headache, and the appetite was returning; he was now beginning to enjoy his breakfast, which he had not done for the last nine months; and he slept a great deal better at night. ℞. Zinc. Ox., gr. iv, bis die, and apply the following lotion to the face: Boracis, Ðj; Aq., ʒviij. On the 3d of January he felt quite well. I then ordered for him a mixture of liquor arsenicalis and carbonate of potash, and he called again for the last time on the 10th of January, when the eruption had nearly disappeared.

W. H— (Case 7), aged 58, an emery sifter; admitted on the 9th of August, 1858.

Upwards of twenty years ago he contracted the habit of drinking to excess. His favorite beverage consisted of gin, which he gradually increased to four glasses a day, and he took from one to three pints of beer daily. He admits that this habit has lasted during fifteen years; but for the last five or six years he has reduced his allowance to an average of two pints of beer daily, and no spirits. When eighteen years old had syphilis, and about twenty years ago was attacked with fits (apparently epileptic), but which have not returned since that time. His present illness began six years ago. He complains of much giddiness, with trembling; sleeps little at night, is then very restless. Frequently fancies he sees cats and rats, has thought there were rats on his bed. States also that on several occasions during the last six years he has rambled a great deal in his mind, and according to his own impression, " People about him have thought he was going mad." Also —muscæ volitantes, and an occasional sensation of choking. Had delirium tremens two years ago, after an accidental indulgence, although he had then given up the habit of drinking to excess. He was first treated with a decoction of aloes, and the hospital mixture of iron and quassia, from which, however, he derived no benefit.

On the 11th of October, two grains of oxide of zinc were prescribed for him, to be taken twice a day, and on the 14th the dose was increased to four grains.

18th.—Since the 14th, feels himself better, sleep much improved, not so giddy; headache and muscæ volitantes diminished; sensation of choking less urgent. Zinc. Ox., gr. vj, bis die.

21st.—Has slept well for the last two or three nights; giddiness and headache considerably less. Has noticed that he feels sleepy in the day, and if he sits down to read, he falls asleep; he was never subject to sleep in the daytime until he took the oxide of zinc; his own words are "I got no sleep, or very little, night or day, before beginning the powders;" he becomes sleepy about an hour after taking the medicine. Complains of sickness at night when he awakes; appetite somewhat improved. Has completely given up every kind of alcoholic liquor, and drinks nothing now but tea and water. I advised him to take half a pint of beer daily. P. pulv., gr. vj, bis die.

24th.—Improvement continues. P. pulv., gr. vj, bis die.

29th.—Improvement remarkable; no longer any uneasy sensation in the throat, very seldom suffers from muscæ volitantes; feels stronger, although still weak. P. pulv., gr. vj, bis die.

November 1st.—Feels quite well, although still weak. The oxide of zinc was now discontinued, and I prescribed for him the hospital mixture of iron and quinine.

8th.—Feels stronger, allows himself one or two pints of ale daily.

The patient now left off attending, but on November the 30th, I called upon him, and found that he had been gradually recovering his strength, that he slept tolerably well, although not so soundly as when under treatment, and considered himself perfectly recovered from his long illness. He took very little beer, say less than one pint daily; felt quite capable of working if he had any work to do.

J. L— (Case 29), aged 24, a butcher; admitted on the 21st of January, 1859.

Has been in the habit, during eight or ten years, of drinking Port wine to the extent of a bottle a day, with a daily allowance of half a pint of spirits, but has seldom been intoxicated. For the last five or six years he has diminished his allowance of stimulants, and during the preceding twelve months has drunk no more than about one pint of beer daily. The first symptoms of his present illness showed themselves eighteen months ago. He suffers from headache, want of sleep, slight trembling in the morning, and perceives, at times, passing shadows. He is very weak, especially in the hips, and complains of pain in the knees.[1] To take two grains of oxide of zinc twice a day.

On the 27th there was already an improvement in his health, and the dose was doubled; on the 31st, I took the following note:—

---

[1] This symptom, although being probably independent of alcoholism, has appeared to me, in the present instance, so closely connected with the disease as not to warrant the case being classed under another head.

Sleeps nearly all night, and is much less troubled with nightmare; no longer any passing shadows; appetite improved. Pain in the knees and weakness in the hips about the same. Has a refreshing sleep of about an hour in the afternoon; feels sick after taking the powders. To rub the knees with turpentine liniment, and to continue taking four grains of oxide of zinc twice a day. On February the 10th, as he stated he had suffered much from headache and giddiness during the preceding week, the dose of oxide of zinc was reduced to two grains,[1] which he continued taking until he left off attending. The 24th of February I reported the following note of his case:—Sleeps very well, appetite very good; is only subject now to a little trembling when taking active exercise, and is then obliged to sit down; is improving in strength; knees still painful. He observes, he has suffered for the last eighteen months

---

[1] This was one of those remarkable cases in which too large doses of oxide of zinc have brought on symptoms resembling those arising from the disease itself; by diminishing the dose these unfavorable symptoms disappeared.

from his present illness, has consulted four medical men, and obtained no relief until he underwent a treatment with oxide of zinc. On that day a small blister was ordered to be applied to the knee. March the 7th.— Strong exercise no longer brings on trembling, is free from every symptom of chronic alcoholism, and only experiences pain in the knee. Is directed to return a fortnight afterwards. He called on the 4th of April, and announced himself quite well, only feeling occasionally somewhat faint when actively at work.

W. J— (Case 5), aged 75, a shoemaker; admitted as out-patient on June the 7th, 1858.

When a young man he became addicted to hard drinking, which he continued indulging in for about twelve years, when, after a sharp attack of delirium tremens, he adopted habits of sobriety, taking no more than about one pint of porter daily, with an occasional glass of spirits. About six weeks previous to admission, he became subject to trembling; since then, complains also of headache, occasional giddiness, and slight hallucinations; lies awake for three or four hours every night, but sleeps very well the remainder of the night. His appetite is bad. Feels weak, and has been obliged to give up work. To take two grains of oxide of zinc twice a day.

By the 14th of June the dose of oxide of zinc had been gradually increased to four grains; on the 21st, is a great deal better, although still weak. Sleep improved; headache diminished; complains of a sensation of nausea occurring about an hour after taking the powders, and feels sleepy at that time.[1] Five grains of oxide of zinc were prescribed, to be taken twice a day. He returned on the 28th, on which day I took the following note. No longer any trembling; sleeps now very well at night; powders do not produce sickness. To continue with five grains of oxide of zinc. On the 1st of July he was quite well in every respect, though still weak; was ordered to take a mixture of iron and quinine. He called for the last time on the 5th of July.

---

[1] A blister had been applied previously, behind the neck, but the date of this prescription has not been recorded in my note-book.

CLASS II.—Division I.

Cases of Chronic Alcoholism, accompanied by other diseases, and occurring during the period of excessive indulgence.

E. B— (Case 9), aged 41, a tailor; admitted on the 18th of October, 1858.

Has been addicted to hard drinking for the last twenty years, and takes on an average four or five pints of porter, and two or three glasses of gin daily; but has drunk repeatedly as much as a gallon of beer in the day. Has been intoxicated occasionally, though not severely; has suffered from ill health for twelve or thirteen years. For the last eight years his appetite has been very deficient, and he feels a constant pain in the stomach, which is increased after meals, and is attended with sickness, especially in the morning. Has had a bad cough for the past three months, complains of shortness of breath, and has suffered from acute pain in the region of the kidneys, which, however, he feels now much less than formerly. Sleep very restless, and disturbed by dreadful dreams. Occasional headache and constant giddiness; muscæ volitantes, and now and then dimness of sight; occasional hallucinations especially at night when falling asleep, and is subject to trembling. On the 25th of October, having derived little or no benefit from the treatment adopted, I prescribed for him two grains of oxide of zinc, to be taken twice a day. This dose was increased to six grains on the 29th. He gave up attending from the 1st of November, and on that day I took the following note of his state: Is now very much troubled by his cough, though he sleeps more comfortably at night. Sickness diminished; still trembles a little, but less than before. Pain in stomach not so acute.

This patient applied again for advice on the 6th January, 1859, having drunk three pints of porter and a glass of spirits daily since he left the hospital. He was again suffering from symptoms of chronic alcoholism, and was treated as before with oxide of zinc. On the 20th, slept well, no more trembling, occasional headache; bad appetite. On the 27th, the appetite was improving; he no longer suffered from giddiness, and the headache was but trifling. He then gave up attending the hospital.

W. B— (Case 10), aged 56, a labourer; admitted on the 2d of October, 1858.

He first contracted the habit of drinking when in France, twenty-six or thirty years ago. Some years later he returned to this country, and worked in the fields as a labourer for nine years, during which time he lived a sober life. He afterwards resided at Peckham, and drove a coal van, when he drank from three to six pints of beer daily, and occasionally a little rum; this he continued doing for two years. During the following nine years he was still employed driving a coal van, although living in another part of the town, and allowed himself the same quantity of beer and spirits. For the next three years he was employed at a wharf, but was engaged more especially in driving a cart; he then diminished his allowance of stimulants to two pints of beer and no spirits. Being employed afterwards at some gas works, he took for two years and a half about three pints of beer daily. During the last two years and a half he has been at work in a brewery, and in the habit of drinking six or seven pints of beer a day, taking but little solid food—being sometimes a whole day without eating anything. He has not often been drunk, and never had delirium tremens. Within the last year has been very subject to giddiness; when reading he sometimes loses his sight entirely for five minutes; he also frequently sees objects double, and is troubled with muscæ volitantes. His wife says he has frequently stated, during his illness, that he saw rats on his bed; hears a sound of bells ringing, especially in the right ear. A fortnight before applying for relief, this patient was seized with a pain in the right lumbar region, which has continued since then, being at times very acute, especially when he is in bed. Has passed blood in his stools several times; about the 11th of October was suffering from this symptom, although to no severe extent. Complains of pain in the stomach, sometimes very great after meals; feels very weak, especially in the left leg; has a cough. W. B— was first treated for bronchitis and rheumatism, but as he did not improve, I was induced to look still more closely into his case, and then found him suffering principally from the effects of long-continued intemperance.

October 28th, was ordered to take two grains of oxide of zinc twice a day. The dose was increased to four grains on the 1st of November, and on the 4th to eight grains. The following notes were taken on the 8th: Trembling much diminished; no longer any hallucinations or muscæ volitantes, or noises in the ears. Still complains of want of sleep. P. gr. viij, bis die. On the 11th, ten grains of oxide of zinc were ordered for a dose, and, as he was complaining of his cough, I prescribed our hospital pectoral mixture. A few days after (date omitted)—Is now disturbed at night by cough, but no longer from uneasy sensations and restlessness, as formerly; has perfectly recovered from the trembling; appetite is excellent. Takes two pints of ale daily, and is in very good spirits. Is cured of chronic alcoholism.

J. H— (Case 25), aged 27, a shoemaker; admitted on December the 9th, 1858. Eighteen months ago contracted the habit of drinking to excess, taking four or five pints of beer and a glass of spirits daily, but during the last twelve months drinks three pints of beer daily, and spirits occasionally. Complains of cough, and of pain in the stomach, especially after eating. Is in very low spirits, and fancies at times that people are wishing to do him some injury. Trembles a great deal, more especially in the evening. Cannot sleep at night; is troubled with ringing in the ears, and black specks flying before his eyes; much weakness in the knees. Has been frequently drunk, but never had delirium tremens. He began suffering from the foregoing symptoms shortly after taking to hard drinking.

This patient was treated at first for bronchitis, with cod-liver oil, and a mixture of iron and quinine; from which treatment he derived little or no benefit. I then discovered he was suffering from the effects of alcoholic stimulants, and prescribed for him, on January 6th, two grains of oxide of zinc to be taken twice a day.

On the 10th of January there was already a marked improvement; the dose was increased to gr. iv. The following note was recorded on the 13th: No trembling, sleeps well, giddiness very slight; complains of cough, especially in the morning; feels much stronger; drinks now very little. I prescribed half an ounce of cod-liver oil to be taken three times a day, and five grains of oxide of zinc to be taken twice a day. On January 17th the dose of oxide of zinc was reduced to three grains, and continued with the cod-liver oil till the 24th of January, when he called at the hospital for the last time. On that day the cough had rather increased, and he occasionally observed muscular twitchings in the fingers and legs. He felt no more, besides these symptoms, than a slight nervous uneasiness, and a shooting pain over the right eyebrow.

CLASS II.—Division II.

Cases of Chronic Alcoholism complicated by other diseases, and occurring after the habit of excessive drinking had been given up.

J. R— (Case 23), aged 34, a policeman; admitted on the 17th of December, 1858.

During a period of two years he drank on an average one gallon of beer and three glasses of spirits daily; but for the last ten years has given up habits of intemperance, taking one or one pint and a half of beer daily, and scarcely any spirits. He has been subject to coughing for several years past, complains of pain in the stomach when fasting; appetite pretty good. Trembles a great deal in the morning; does not sleep well, and dreams a great deal; slight hallucinations at night, or when he shuts his eyes; weakness in legs excessive. Has never been quite well for the last ten years, but only gave up work on the 2d of November.

The patient was first treated with cod-liver oil and quinine, from which he derived no benefit. On January 3d, I prescribed for him two grains of oxide of zinc to be taken twice a day. On the 6th the dose was increased to four grains, and on the 10th to six grains. Was then better; trembling diminished, sleep improved, in better spirits, stronger, no longer any hallucinations; he complained of pain in the region of the right kidney, was directed to rub this part with compound soap liniment. On the 17th of January the dose of oxide of zinc was reduced to four grains. He applied for the last time on the 20th, the pains in the loins and stomach had somewhat diminished; felt quite well in other respects.

D. B— (Case 27), aged 72, a greengrocer;[1] admitted on the 17th of January, 1859.

This patient, it may be remembered, was suffering from symptoms of chronic alcoholism complicated by an attack of gout. He had, from youth, taken to the habit of drinking, and been subject to gout from the age of

[1] For the description of this case, see p. 50.

sixteen. The symptoms of chronic alcoholism diminish or almost entirely leave him during the intervals between the attacks of gout, and increase on the return of this illness, the intermittent disorder of the nervous system having occurred, as far as could be ascertained, since the age of nineteen. On the day of his admission I prescribed for him two grains of oxide of zinc, to be taken twice a day, which dose was increased to gr. iij, on the 20th, when the nervous symptoms had already subsided. On the 24th, I prescribed, in addition to gr. v of oxide of zinc, a mixture of carbonate of potash and rhubarb. On the 27th the powders were omitted, and ℥x of colchicum wine were added to the mixture, which was continued; on the 31st, was directed to rub the painful parts with compound soap liniment, and take three grains of compound soap pill every night. On the 3d of February the attack of gout was decidedly abating, and the patient had altogether improved much beyond his anticipations. The same treatment was continued, and on the 14th of February the dose of compound soap pill was increased to gr. v. The 24th of February, the attack of gout had passed off. I ordered half an ounce of codliver oil to be taken three times a day, and two grains of oxide of zinc twice a day. On March 3d, another attack of gout was impending, there being a return of pain under the feet and swelling in the left thumb; but no relapse of the nervous symptoms occurred. I had again recourse to the mixture of carbonate of potash and rhubarb, to which $\mathfrak{m}x$ of colchicum wine were subsequently added, and omitted the oxide of zinc. The attack of gout ran a mild course, and without any recurrence of the affection of the nervous system. On the 14th the attack of gout was mitigated, and on the 24th it had passed off, leaving the thumb of the left hand slightly swollen, and the great toe disposed to swell towards the evening and occasionally painful at night. The patient was then discharged, apparently cured of the affection of the nervous system, though still predisposed to gout.

OBSERVATIONS ON SYNOPTICAL TABLE.

Having glanced at the synoptical table, the reader will probably ask whether these cases include all the patients suffering from chronic alcoholism, who have been placed under my care at the Westminster Hospital. The answer to this inquiry is, that with the exception of one or two patients, not reported because the notes of their illness were not deemed sufficiently complete, the synoptical table is to be considered as offering a faithful account of the whole of the cases of chronic alcoholism that have come under observation in my hospital practice, from a short time after my being appointed to the Westminster Hospital. I must, however, remark that the number of cases described falls short to a considerable extent of the real number of hospital patients suffering from chronic alcoholism who have applied to me for medical advice; as, previous to my attention being called to the disease in question, I must have frequently overlooked the influence of spirituous liquors on the nervous system; an omission partly owing to the circumstances alluded to in the beginning of this work, and partly to the fact that, in many instances, the prominent symptoms depended on morbid conditions which differed widely from those known to be caused by the abuse of alcoholic beverages.

Turning our attention to the employments of the patients, we observe that of the forty-eight cases reported, there are five whose occupations are unknown; those of the other forty-three are as follows:

Labourers	8
Coal porters	3
Carpenters (including chairmaker)	3
Shoemakers	3
Engineers	2
Hawkers	2
Cabmen	2
Sailor	1
Emery-sifter	1
Tailor	1
Cutler	1
Carrier	1
Carter	1

Grocer	1
Greengrocer	1
Painter	1
General dealer	1
Cushion-maker	1
Workman in a brewery	1
Policeman	1
Omnibus driver	1
Butcher	1
Clerk	1
Stoker	1
Weaver	1
Gas fitter	1
Check-taker	1
	43

Consequently the most numerous class of patients are, with the exception of shoemakers, those employed in works which require great muscular exertion in the open air, and who have recourse to drink in order to stimulate their overtasked strength, these are: eight labourers, three coal porters, and three carpenters. We next observe two hawkers, two engineers working steam-engines, and two cabmen. Hawkers drink with the idea that it assists them to get over their daily muscular and vocal exertions; engineers may thus hope to make up for the loss of strength arising from excessive perspiration; and cabmen, having often nothing to do, spend their time and money in the tap-room, leaving their cab in the rank. The other classes of individuals included in the table have each furnished one instance of chronic alcoholism.

I have not much to add in connexion with the quantity and quality of the spirituous beverages taken. The reader must recollect that the information contained under this head must be taken with reserve. Some patients stated they had been intoxicated regularly every day; others only once a week —usually on Saturday evening; others drank to a sufficient extent to bring on an attack of chronic alcoholism, although they only indulged to excess at irregular intervals; others again would perhaps have escaped the disease, had they not lived too freely at Christmas. Thus it will be perceived there is much difficulty in placing in a tabular form the amount and nature of alcoholic beverage taken by each individual.

The period during which habits of intemperance existed was equally difficult to ascertain with satisfactory correctness, for it often occurred that the allowance of beer and spirits had been *gradually* increased; and it was impossible to arrive exactly at the date when the quantity taken had reached such an extent as to interfere with health. It also frequently happened that an individual, having drunk freely for a certain number of years, suddenly reduced his allowance within moderate limits, and then again took to intemperate habits; this could not well be expressed in a synoptical table.

Of the forty-five cases where the condition of sleep was recorded, in one only had this function not been affected, and in the present instance (Case 48) I am disposed to believe that the individual suffered as much, if not more, from smoking, as from drinking. Sleep, disturbed by excesses in spirituous beverages, has this peculiarity—that it is always attended with restlessness and mental anxiety. I do not remember hearing a patient suffering from chronic alcoholism state that he lay in bed comfortably, although he could not sleep. This symptom varied considerably in degree; in many cases, individuals were but slightly affected, frequently awaking from a sleep disturbed by nightmare; other patients, more seriously attacked, not sleeping at all, and troubled all night with frightful hallucinations.

Respecting the other symptoms of chronic alcoholism, I have merely to call the reader's attention to the fact that *hallucinations*, *trembling*, and *giddiness*, are, with but few exceptions, constantly present; and these symptoms, when occurring in conjunction with sleeplessness, appear to me as all but characteristic of chronic alcoholism.

The column headed *symptoms not referrible to the nervous system*, shows how frequently chronic alcoholism is attended with other diseases, more especially affections of the digestive organs; these secondary disorders being frequently, in my opinion, the immediate cause of the attack of chronic alcoholism. The next column, headed *pre-existing disease*, indicates that chronic alcoholism is very frequently preceded by one or more attacks of delirium tremens, and also that the patient may suffer previously from chronic alcoholic intoxication; it will be perceived, however, that the illness occurs not unfrequently as the first indication of a disordered constitution from frequent indulgence in spirituous drinks.

Finally, with respect to the treatment, it must be remembered that nothing is so difficult as to keep up a regular series of observations on hospital out-patients; for some apply only once for relief; others give up attending as soon as they feel themselves well enough to return to work; and a few, only, can be made to understand the importance of calling regularly at the hospital until they are discharged. It will be satisfactory, however, to state that out of the forty-eight patients who were treated for chronic alcoholism, only six returned for advice on a recurrence of the unfavorable symptoms.

OBSERVATIONS ON SYNOPTICAL TABLE. 147

I have endeavoured in this treatise, to bring prominently to light the most frequent form of disease produced by excessive drinking; and, although it has been considered advisable to confine the subject strictly within medical limits, it will be a source of satisfaction to me should any of the foregoing observations tend to strengthen the hands of those philanthropists who have devoted their time and exertions to the repression of excesses in alcoholic stimulants.

SYNOPTICAL

GIVING AN

FORTY-EIGHT CASES OF

Treated by the Author, at

No.	Name and Age of Patient. Profession.	Quality and Quantity of Spirituous Beverage taken.	Period of existence of Intemperance.	Sleep (state of).	Hallucinations.
1	W.B., æt. 34, carpenter	One pint of gin daily; and two or three, and occasionally six or eight pints of beer	Unknown	No sleep	Unknown
2	C.A., æt. 33, profession unknown	Accustomed to hard drinking, but quality and quantity unknown	Unknown, but turned teetotaller seven weeks ago	Unknown	Unknown
3	T.S., æt. 33, engineer on board a steam-boat	About one pint of spirits, and four or five pints of beer daily	Unknown	Want of sleep	Unknown
4	G.P., æt. 35, profession unknown	A great deal of brandy and water, and occasionally pure brandy, and about eight pints of beer daily	Unknown, but began at least six years ago	Cannot sleep	Hallucinations

TABLE,

ACCOUNT OF

CHRONIC ALCOHOLISM,

the Westminster Hospital.

Trembling.	Other Symptoms referrible to the Nervous System.	Symptoms not referrible to the Nervous System.	Pre-existing Disease.	Result of Treatment.
Unknown	Giddiness	Great weakness and spasmodic dyspnœa	A slight attack of delirium tremens	Cured, but will not give up drinking, and applies again subsequently.
Unknown	Unknown	Palpitations and occasional pain in the heart, slight bruit at first sound	Unknown	Dicharged cured.
Tongue slightly tremulous	Giddiness	Thirst, pain in epigastrium, spasmodic dyspnœa.	Several attacks of delirium tremens	Discharged cured.
Fits of violent trembling	Dizziness of sight	Appetite bad; occasional pain in region of the bladder; urine turbid and reddish.	Two slight attacks of delirium tremens within the last two years	Ceased attending, having quite recovered.

No.	Name and Age of Patient. Profession.	Quality and Quantity of Spirituous Beverage taken.	Period of existence of Intemperance.	Sleep (state of).	Hallucinations.
5	W. J., æt. 75, shoemaker	Excessive use of alcoholic liquors, quality and quantity unknown	From youth until twelve years ago	Does not sleep well	Passing shadows
6	E. B., æt. 40, sailor	Began with twelve glasses of brandy daily, and afterwards took six glasses of rum and five or six pints of beer occasionally	Ten or eleven years ago	Does not sleep well	Passing shadows
7	W. H., æt. 58, emery-sifter	Gin, gradually increased to four glasses a day, and from one to three pints of beer daily	More than twenty years, but for the last six years has reduced his allowance	Sleeps seldom at night	Hallucinations
8	J. P., æt. 39, coal-porter	Five pints of beer and a glass of gin daily	Twenty years	Sleep very restless	Occasional passing clouds
9	E. B., æt. 41, tailor	Four or five pints of porter and three glasses of gin daily	Twenty years	Cannot sleep	Occasional hallucinations and muscæ volitantes

TABLE. 151

Trembling.	Other Symptoms, referrible to the Nervous System.	Symptoms not referrible to the Nervous System.	Pre-existing Disease.	Result of Treatment.
Trembling for the last six weeks	Giddiness and headache	Does not micturate freely; water high-coloured and turbid	An attack of delirium tremens twelve years ago	Ceased attending, having quite recovered.
Legs tremble occasionally	Occasional giddiness, headache, and tinnitus aurium	Legs very weak, especially the right	An attack of delirium tremens in 1843; also fits (probably epileptic) when 20 years old	Discharged cured; slight weakness remains.
Trembling	Giddiness	Unknown	Syphilis at æt. 18; at æt. 38 fits (probably epileptic); an attack of delirium tremens two years ago	Discharged cured.
Legs tremble when he sits	Giddiness	Knees very weak; acute pain in right hip, and occasionally shooting pain across the temples	Unknown	Ceased attending, having quite recovered.
Trembling	Headache and much giddiness; tinnitus aurium	Pain in epigastrium; bad appetite, palpitations, cough, shortness of breath	Unknown	Ceased attending, having much improved.

No.	Name and Age of Patient. Profession.	Quality and Quantity of Spirituous Beverage taken.	Period of existence of Intemperance.	Sleep (state of).	Hallucinations.
10	W. B., æt. 56, labourer	Three to six pints of beer daily, and a little rum occasionally	Twenty-six years	Does not sleep well	Muscæ volitantes and hallucinations
11	J. W., æt. 18, hawker	Occasionally from seven to eight pints of beer and a glass of gin	Five years	Very restless at night, and does not sleep well	None
12	C. M., æt. 27, carrier	Addicted to the excessive use of alcoholic beverages, quality and quantity unknown	Unknown	Very restless at night, and does not sleep well	Passing shadows
13	J. T., æt. 55, labourer	Three or four pots of beer daily, and two glasses of gin	Eight or nine years, but rather less for the last five years	Does not sleep at all well at night	None
14	J. I., æt. 39, labourer	Three or four pints of beer daily, no spirits	Unknown	Restless at night, and does not sleep well	Passing clouds and muscæ volitantes
15	E. C., æt. 49, grocer	Half a pint of brandy and five or six pints of stout daily	Eight years, but altered his allowance lately to eight pints of porter a day	Cannot sleep well	None but muscæ volitantes

Trembling.	Other Symptoms referrible to the Nervous System.	Symptoms not referrible to the Nervous System.	Pre-existing Disease.	Result of Treatment.
Trembling of legs and arms	Vertigo, transient blindness	On admission bronchitis and rheumatism; has passed blood in stools	Never had delirium tremens	Discharged cured.
Occasional trembling of legs	Giddiness	Nausea and great weakness	Repeated attacks of drunkenness	Discharged cured.
Trembling	Diminished intellect, and fear of being injured	Pains in both hips, and legs weak	Unknown	Ceased attending, having much improved.
None	Great giddiness	Pain in umbilical region after taking food	Right arm paralysed thirteen years ago	Discharged cured.
Occasional trembling	Tinnitus aurium	Pain in the stomach; coughs	Unknown	Ceased attending, having much improved.
Trembles every morning when getting up	Giddiness	Sick in the morning, and choking sensation	A slight attack of delirium tremens two months ago	Discharged cured.

No.	Name and Age of Patient. Profession.	Quality and Quantity of Spirituous Beverage taken.	Period of existence of Intemperance.	Sleep (state of).	Hallucinations.
16	S. H., æt. 41, carpenter	Four or five pints of porter and a quarter of spirits daily	From youth; about a year ago reduced his allowance to three pints of beer and one glass of spirits	Does not sleep at all at night	Hallucinations
17	L. M., æt. 46, painter	Three glasses of spirits and three pints of ale daily	From youth until the last three years; reduced, at that time his allowance to three pints of beer and one glass of spirits	Restless at night, and does not sleep well	Passing clouds
18	G. R., æt. 38, general dealer	From three glasses of spirits to a pint daily, and four or five pints of ale	From youth to about eleven months ago, when he gave up beer and spirits	Cannot sleep at all at night	Passing clouds
19	G. M., æt. 33, cushion-maker	Six pints of beer daily	For the last twelve years	Unknown	Passing shadows

Trembling.	Other Symptoms referrible to the Nervous System.	Symptoms not referrible to the Nervous System.	Pre-existing Disease.	Result of Treatment.
Occasionally violent trembling	Much giddiness and headache	Pain in epigastric region; morning sickness	None	Discharged cured.
No trembling, but occasional starting	Much giddiness and headache	Pain in the stomach after eating; an attack of hæmoptysis	Had delirium tremens twice three years ago	Only attended twice.
Slight trembling	Much headache and giddiness; and transient blindness	Appetite very bad, but no other dyspeptic symptom; eczema on face	None	Cured.
Much trembling at times	Tinnitus aurium and giddiness	Weakness, shooting pains in left eye	Unknown	Ceased attending, having much improved.

No.	Name and Age of Patient. Profession.	Quality and Quantity of Spirituous Beverage taken.	Period of existence of Intemperance.	Sleep (state of).	Hallucinations.
20	J. H., æt. 54, chair-maker	Four or five pints of beer and three glasses of spirits daily	Ten years; but for the last four years takes two pints of beer and one glass spirits	Does not sleep well	Occasionally passing clouds
21	C. P., æt. 54, labourer	Twelve pints of beer daily	Nine years; but for the last four years takes two or three pints of ale daily	Cannot sleep	None
22	J. L., æt. 53, workman in a brewery	Three or four pots of ale and two or three glasses of gin daily	From youth; but for the last eight or nine years takes three pints of beer and one or two glasses of gin daily	Does not sleep at all well	None
23	J. R., æt. 34, policeman	One gallon of beer and three glasses of spirits daily	Two years, but has been very moderate for the last ten years	Cannot sleep at all well at night	Slight hallucinations at night or when eyes are shut

TABLE. 157

Trembling.	Other Symptoms referrible to the Nervous System.	Symptoms not referrible to the Nervous System.	Pre-existing Disease.	Result of Treatment.
Unknown	Headache	Bronchitis	Bronchitis every winter, never had delirium tremens	Cured of alcoholism, remains under treatment for bronchitis.
Trembling	Transient loss of sight	Unknown	None	Has not attended after first visit.
Unknown	Slight headache	Knawing pains in legs, and such weakness that he can hardly walk	None	Left off attending much relieved; legs continue very weak.
Trembles a great deal, in the morning especially	None	Legs very weak, and bronchitis	Subject to bronchitis; had delirium tremens twice during the two years of intemperance	Left off attending, having quite recovered from chronic alcoholism, but still suffering from pain in the back and neck.

158   SYNOPTICAL

No.	Name and Age of Patient. Profession.	Quality and Quantity of Spirituous Beverage taken.	Period of existence of Intemperance.	Sleep (state of).	Hallucinations.
24	W. F., æt. 34, carter	Three or four pints of beer and three or four glasses of spirits daily	Began when young, but subsequently reduced his allowance	Does not sleep well	Hallucinations
25	J. H., æt. 27, shoemaker	Four or five pints of beer and one glass of spirits	Began twelve months ago, but for the last six months diminished allowance	Cannot sleep at all	Muscæ volitantes
26	W. P., æt. 29, cabman	Three or four glasses of gin and two or three glasses of beer daily	Nine years	Cannot sleep well	Hallucinations and passing clouds
27	D. B., æt. 72, greengrocer	Four or five glasses of spirits and three or four pints of beer daily	From youth to fourteen or fifteen years ago, when reduced allowance	Very little sleep	Passing fog and hallucinations
8	D. W., æt. 56, engineer	A quarter of a pint of gin daily, and half a pint of whisky once a week	Unknown, but drinks less than previously for the last two years	Sleep very restless	None

TABLE.

Trembling.	Other Symptoms referrible to the Nervous System.	Symptoms not referrible to the Nervous System.	Pre-existing Disease.	Result of Treatment.
Knees tremble	Giddiness	Pain in the stomach, weakness in knees; slight pain in hips	None	Ceased attending, having much improved.
Much trembling, especially in the evening	Tinnitus aurium and giddiness	Pain in the stomach; choking sensation; occasional weakness in legs; coughs	Unknown	Gave up attending, having quite recovered, but still subject to slight general nervous uneasiness.
Trembling, especially in the morning	Giddiness	Choking sensation; weakness in legs and across the hips	Subject to bronchitis, otherwise in good health	Discharged cured, but slight dry cough remaining.
Trembles a great deal in the morning	None	Frequent sickness and vomiting; an evident attack of gout	Subject to gout	Discharged cured of alcoholism, though not of the gouty predisposition.
Trembles in the morning	Dimness of sight and headache	Weakness and sickness in the morning	None	Ceased attending, greatly relieved.

No.	Name and Age of Patient. Profession.	Quality and Quantity of Spirituous Beverage taken.	Period of existence of Intemperance.	Sleep (state of).	Hallucinations.
29	J. L., æt. 24, butcher	A bottle of port wine daily, and half a pint of spirits	Twelve or fifteen years, but diminished allowance for the last four or five years, and during the last year, one pint of beer daily	Sleep greatly disturbed by horrid dreams	Occasional passing shadows
30	J. H., æt. 49, no profession, but has been policeman	Daily average three or four glasses of spirits and two pints of ale	Twenty years; but for the last twelve months, half a pint of beer daily	Sleeps well, but dreams a great deal	None
31	J. B., æt. 40, labourer	About five quarts of beer and a pint of gin daily	Unknown	Very little sleep and dreadful dreams	Unknown
32	W. D., æt. 34, profession unknown	Daily average a quart of beer and about one pint of gin with peppermint	From about nine years ago to five years ago, when he left off drinking	Sleep much disturbed	Passing shadows and hallucinations

TABLE.

Trembling.	Other Symptoms referrible to the Nervous System.	Symptoms not referrible to the Nervous System.	Pre-existing Disease.	Result of Treatment.
Trembles a little in the morning	Headache	Weakness in the hip	None	Dismissed cured.
Tongue slightly tremulous	Headache and much giddiness, dimness of sight and tinnitus aurium	Pain in right hypochondriac region; no other symptoms	Had an attack of the same symptoms four years ago, and was treated at St. Thomas's Hospital	Dismissed cured.
Trembles in the morning	Much headache and giddiness, muscæ volitantes, is very low spirited	Pain in the stomach; no appetite; pain in dorsal spinal region; great weakness	Several attacks of delirium tremens; has been subject to cough	Ceased attending, much relieved.
Trembling	Headache and much giddiness, with faintness; noises in the ears	Coughs; pain in chest; sickness after taking food	Unknown	Ceased attending, much relieved.

11

No.	Name and Age of Patient. Profession.	Quality and Quantity of Spirituous Beverage taken.	Period of existence of Intemperance.	Sleep (state of).	Hallucinations.
33	H.E., æt. 24, clerk	Has been drunk five or six times, was so last Christmas; is very moderate between excesses	Unknown	Does not sleep well	None
34	T. H., æt. 36, cutler	Sixteen or seventeen wine-glasses of rum daily; occasionally brandy instead of rum	Fourteen months. During the last seventeen years has diminished his allowance, being occasionally intoxicated.	Awakes often at night, but sleeps pretty well	Unknown
35	G. B., æt. 28, stoker in House of Parliament	One pint of ale daily and no spirits	Unknown, but became a teetotaller three years ago	Sleeps very indifferently	Hallucinations
36	B. L., æt. 29, shoemaker	On an average two quarterns of rum and three or four pints of beer daily	For the last two years, but took to drinking five years ago	Sleeps pretty well, but sleep disturbed by dreams	Hallucinations

Trembling.	Other Symptoms referrible to the Nervous System.	Symptoms not referrible to the Nervous System.	Pre-existing Disease.	Result of Treatment.
Trembles a great deal in the morning	No headache no giddiness	Much weakness	Unknown	Dismissed much relieved.
Trembles in the morning	Much headache and occasionally much giddiness	Pain in the back when walking	Has suffered from the same symptoms twice previously; the first time six or seven years ago, after being garotted	Ceased attending, having quite recovered; still complains of pain in the back.
Trembles in the morning	Giddiness, transient blindness, headache, muscæ volitantes	Rheumatic pains; no dyspeptic symptoms	Unknown	Only attended one day at the hospital.
Trembles in the morning	Headache and giddiness	Appetite deficient, much pain after eating in epigastric region, pain in dorsal region.	Has attended the hospital two years ago for the same illness	Dismissed cured.

No.	Name and Age of Patient. Profession.	Quality and Quantity of Spirituous Beverage taken.	Period of existence of Intemperance.	Sleep (state of).	Hallucinations.
37	T. R., æt. 48, weaver	From four to ten pints of porter daily; no spirits	For twenty years, but diminished allowance for the last two years	Sleeps but very little; is much troubled with nightmare	Hallucinations
38	J. H., æt. 56, labourer	Daily average, five or six pints of beer and two or three and a half quarterns of gin	From three years ago till the middle of last year, but drank hard though not so much, previous to that time	Sleep much interrupted, and restless	Slight hallucinations
39	B. M., æt. 44, omnibus-driver	Four or five pints of porter and two or three glasses of spirits daily	For twenty years	Restless at night; frequently awakes dreaming	Slight hallucinations
40	W. B., æt. 49, coal-porter	Five or six pints of beer and a glass of spirits daily	Twenty years	Cannot sleep well at night, but not subject to nightmare	None

TABLE.

Trembling.	Other Symptoms referrible to the Nervous System.	Symptoms not referrible to the Nervous System.	Pre-existing Disease.	Result of Treatment.
Trembles a great deal	Headache	Small tumour on upper part of the sternum, very painful on pressure.	Unknown	Has not attended beyond the first visit.
Unknown	Dizziness and dimness of sight	Pain in the back, much emaciated; case of enlarged liver	Unknown	Gave up attending; sinking from organic disease.
Unknown	Headache and giddiness; and dimness of sight, muscæ volitantes	Coughs and spits a great deal, pain in stomach after eating, frequent sickness in the morning	Great mental depression from family afflictions	Ceased attending, much relieved.
Trembling	Unknown	Cough, sickness, pain in stomach, and bad appetite	None	Ceased attending, much relieved.

No.	Name and Age of Patient. Profession.	Quality and Quantity of Spirituous Beverage taken.	Period of existence of Intemperance.	Sleep (state of).	Hallucinations.
41	A. P., æt. 40, gas-fitter	Six or seven pints of beer daily	Two or three years	Want of sleep	None
42	G. J., æt. 27, check-taker at a theatre	Four or five glasses of spirits and two or three glasses of ale daily	From youth up to three years ago, and then diminished his allowance to one pint of beer and three glasses of spirits	Cannot sleep well at night, has frightful dreams	Unknown
43	T. C., æt. 48, coal-porter	Two pots of beer daily and a little spirits occasionally	For twenty-four years, but rather more during the last two or three months	Cannot sleep at night	None
44	R. S., æt. 32, labourer	Six or seven pints of ale and two or three glasses of spirits daily	For fourteen years up to a month before admission	Does not sleep well	Hallucinations

TABLE. 167

Trembling.	Other Symptoms referrible to the Nervous System.	Symptoms not referrible to the Nervous System.	Pre-existing Disease.	Result of Treatment.
Trembles in the morning	Giddiness	Six or seven months ago fell from a height of twenty-five feet; no appetite, morning sickness, gastric pain	In good health previous to the accident	Ceased attending, much relieved.
Occasional trembling	Giddiness, headache, transient loss of sight	Pain in the stomach, subject to occasional rheumatic pains	Three or four years ago suffered from coup-de-soleil	Ceased attending, much relieved.
Trembling in the morning when takes more than usual the day before	Muscæ volitantes	A co-existing attack of gout, digestion good	None	Has not attended beyond the second visit.
Trembling, especially in the morning	Giddiness	Morning sickness, but no gastric pain	None	Discharged cured.

No.	Name and Age of Patient. Profession.	Quality and Quantity of Spirituous Beverage taken.	Period of existence of Intemperance.	Sleep (state of).	Hallucinations.
45	A. T., æt. 25, labourer	One pint of beer and three or four glasses gin daily, except on Saturday; this day he takes seven or eight pints of beer and five or six glasses of gin	For seven years	Cannot sleep the four nights following the Saturday excess; sleeps pretty well on the other nights	Hallucinations
46	H. H., æt. 42, hawker	Three or four pints of raw rum daily, and three or four pints of beer	For about twenty-seven years till the last three or four months	Cannot sleep	No positive hallucinations
47	T. D., æt. 42, cab-driver	Began with five or six pints of ale daily, and increased his allowance to twenty glasses of spirits and five pints of ale daily	For twenty-five years	Unknown	Hallucinations
48	W. P., age unknown, occupation unknown	Two or three pints of ale and a glass of spirits daily	For ten or eleven years	Sleeps very well	Unknown

TABLE.

Trembling.	Other Symptoms referrible to the Nervous System.	Symptoms not referrible to the Nervous System.	Pre-existing Disease.	Result of Treatment.
Trembling	Muscæ volitantes, headache and giddiness	Morning sickness and gastric pains	None	Left off attending, having nearly perfectly recovered.
Trembling	Sensation of suffocation, sometimes very acute; giddiness, muscæ volitantes, cramps	Gastric pains, morning sickness	Unknown	Became an in-patient after second visit.
Unknown	Sensation of suffocation	Gout and hæmatemesis; coughs	Several attacks of delirium tremens	Left off attending, quite recovered from chronic alcoholism.
Trembling in the morning	Giddiness and headache	Sickness	Unknown (smokes, and has suffered from it)	Has not attended beyond the first visit.

APPENDIX.

The author begs to observe that the statements contained at pp. 44 and 73 of the present treatise, respecting the property of alcohol, tea, and coffee of diminishing the waste of the body, are derived from the researches of Prout,[1] Dr. Fyfe,[2] Vierordt,[3] Böcker,[4] and Lehmann.[5]

Prout, Dr. A. Fyfe, Vierordt, and Böcker, have found that alcohol, when taken, diminishes the amount of *carbonic acid* expired. Moreover, Böcker informs us that the excretion of *urea* is lessened by alcohol and tea, tea likewise reducing the quantity of carbonic acid evolved.

---

[1] 'Thompson's Annals of Physiology,' vol. ii.
[2] See a paper of Prout, in 'Thompson's Annals of Physiology,' vol. iv.
[3] 'Physiologie des Athmens.'
[4] 'British and Foreign Medico-Chirurgical Review,' vol. xiv, 1854.
[5] See the account of 'Böcker's Researches,' in the last-mentioned periodical.

According to Lehmann's inquiries, the use of coffee also lessens the amount of carbonic acid expired.

The results obtained by Dr. Edward Smith,[1] from recent and very interesting investigations on the action of food upon respiration, agree but partly with the above-mentioned conclusions. This gentleman found that spirits of wine, ale, and stout, increased the quantity of *carbonic acid* expired. Rum commonly increased it, and sherry wine exerted, to a slight extent, a similar action. The amount of carbonic acid evolved, when brandy and gin were taken, was constantly diminished. Whisky varied in its effects. The inhalation of the volatile elements of wine and spirits, lessened the evolution of this gas. Dr. E. Smith also observed that tea and coffee increased the production of carbonic acid, tea being more powerful than coffee.

[1] 'Proceedings of the Royal Society,' 1859, vol. ix, p. 638.

END.

J. E. ADLARD, PRINTER, BARTHOLOMEW CLOSE.

www.ingramcontent.com/pod-product-compliance
Lightning Source LLC
Chambersburg PA
CBHW031444160426
43195CB00010BB/842